Emerging Non-thermal Food Processing Technologies

Basic Text for College Students

Perkins Muredzi

School of Industrial Sciences & Technology;
Harare Institute of Technology Harare, Zimbabwe

CBH
Books

www.cbhbooks.com

Editors: Ernesto Sánchez, and Michael Bergerac
Managing Editor: Manuel Aleman
Designer: Jun Luis Montero

Published in the United States by CBH Books.
CBH Books is a division of Cambridge BrickHouse, Inc.

Cambridge BrickHouse, Inc.
60 Island Street
Lawrence, MA 01840
USA

Library of Congress Control Number: 2012954021
ISBN 978-1-59835-327-3

First Edition
Printed in USA
10 9 8 7 6 5 4 3 2 1

Dedicated to

Mrs. Dorothy Stembile Muredzi,
Sean Muredzi,
and Sheldean Muredzi

Epigraph

Men often become what they believe themselves to be. If I believe I cannot do something, it makes me incapable of doing it. But when I believe I can, then I acquire the ability to do it even if I did not have it in the beginning.

– Mahatma Gandhi

Contents

Chapter 2
Pulsed Electric Field Processing

Chapter 3
Food Irradiation

Chapter 4
Pulsed Light Processing Technology

Chapter 5
Ultrasound Processing Technology

Chapter 6
Oscillating Magnetic Fields Processing Technology

Chapter 7
Ozone

Chapter 8
Microwave and Radio Frequency Processing

Chapter 9
Gas, Cold Plasma Technology

Chapter 10
Overarching Principles: Kinetics and Pathogens of Concern for All Alternative Food Processing Technologies

Preface

The author has observed that there is critical need for literature on non-thermal food processing technologies for freshmen and other college students studying food processing technology at undergraduate and postgraduate levels, and a dire need for information by food industry players and stakeholders on basic information on principles and a general understanding of the technologies now being introduced into the industry.

Statistics show that seven in every 10 people employed worldwide are employed in the food industry and that graduates of food processing technology degree programmes now must have the ability to appraise critically the complex factors, quality, and acceptability of foods produced in an industrial society

Thermal processes such as blanching, pasteurisation or heat sterilisation have long been employed in practice as economical, efficient, reliable, and safe food preservation methods. However, in most cases thermal energy induces various chemical reactions, leading to quality deterioration in certain foods by producing undesirable changes in sensory and nutritional qualities. Thus, preservation technologies that prolong shelf-life without these detrimental effects are to be favoured. Increasing consumer demand for new products with inherent sensory and nutritional qualities has led to the development of gently (minimally) processed foods.

Consumer interest in gently processed foods springs from concerns associated with the fear of chemicals in the food supply, conflicting reports regarding the pros and cons of existing and new food processing technologies, as well as increased awareness of nutritional issues. This consumer attitude has spurred a search for new alternatives to processed foods. non-thermal methods provide such an option because they offer fresh-like 'minimally' processed foods with little loss of colour, flavour, and nutrients. There are several new non-thermal technologies of potential interest to the industry, including high-pressure processing (HPP), pulsed electric fields (PEF), high intensity pulsed light, irradiation, ultrasound cold plasma, ozone, and modified atmosphere packaging.

Among the novel non-thermal technologies, the application of HPP, PEF, and ultrasound to food processing and preservation has become of considerable interest in food research and development. These technologies are of specific interest to the food industry because they not only provide attractive alternatives to conventional methods of thermal processing, which often produce undesirable changes in foods, affecting the balance between high quality and safety, but also offer opportunities for creating new ingredients and products because of the specific actions on biological materials and food constituents The potential to commercialise the non-thermal HPP and PEF technologies as new methods to preserve food products has caught the attention of a food industry wanting to satisfy the consumer's demands for fresh products. With diet and health firmly established as a long-term consumer lifestyle, it is important to understand the impact of the new technologies on the nutritional value of foods.

This text attempts to give an introduction to the various emerging non thermal food processing technologies being considered by the food industry, the basic underlying principles behind each technology and applicability.

Chapter 1

High Pressure Processing Technology

Historical Perspective

The first experiment in food science and technology, the most important work involving microbial inactivation, was reported at the end of the 19th century by Hite (1899), and effects of pressure on physical properties of foods were reported soon after, by others such as Bridgman (1914) (on the coagulation of egg albumin); Payens and Hermens (1969) (effects of the pressure on the beta-casein from milk), and Macfarle (1973) (on pressure–tenderization of meat). After a period of lower research in this field, it has received considerable attention since it was rediscovered approximately 20 years ago.

HPP has had application for years in other industries which process or use ceramics, carbon graphite, diamond, steel/alloy, and plastics. High pressure treatment to kill bacteria was first described in 1895 by Roger. However, the most important work involving microbial inactivation in food science and technology was published in June of 1899 by Bert Hite. Hite reported the effects of HPP on "food" microorganisms by subjecting the milk to pressures of 650 Mega Pascal (MPa) and obtaining a reduction in the viable numbers of microbes. Hite and his co-workers of the University of West Virginia examined a wide range of foods and beverages for the potential use of high pressure processing. Although the concept of using HPP as a method of food processing is not new, the technical impossibility of working at industrial level reduced interest in such a treatment. However, for the last 15 years, the use of HPP has been explored extensively in food industry and related research institutions due to the increased demand by consumers for improved nutritional and sensory characteristics of food without loss of "fresh" taste. In April 23, 1990, the first high pressure products, three kinds of jam (strawberries, kiwifruit, and apple) made using high pressure treatment without application of heat, were introduced to the Japanese retail market. In 1991, yogurts, fruit jellies, salad dressings, and fruit sauces were also introduced and two Japanese fruit juice processors installed semi-continuous high pressure equipment for citrus juice bulk processing.

Introduction and Overview

Increasing consumer demand for minimally processed, additive-free, shelf-stable products, prompted food scientists to explore other physical preservation methods as alternative to traditional treatments such as freezing, canning or drying that rely on heating or cooling operations. Although these technologies have helped to ensure a high level of food safety, the heating, and cooling of foods may contribute to the degradation of various food quality attributes. The colour, flavour, and texture of foods processed solely by heating may be irreversibly altered. To ameliorate the undesirable thermal effects on foods, considerable efforts have been made in commercial and academic circles to develop non-thermal technologies. Investigated technologies are ionization radiation (gamma irradiation), high hydrostatic pressure (HHP), pulsed electrical fields, high pressure homogenization, UV decontamination, pulsed high intensity light, high intensity laser, pulsed white light, mano-thermo-sonication (combined ultrasonic, heat, and pressure), oscillating magnetic fields, high voltage arc discharge, and streamer plasma.

Among these emerging technologies, the most promising ones for food application are high-pressure processing. High-pressure food-preservation and processing technology is being developed to a large extent in reaction to consumers' requirements for foods that are nutritionally healthier, more convenient to use (e.g. easier to store and prepare), fresher (e.g. chill-stored), more natural and therefore less heavily processed (e.g. mildly heated), less heavily preserved (e.g. less acid, salt, sugar) and less reliant on additive preservatives (e.g. sulphite, nitrite, benzoate, sorbate) than previously. Research into the application of HHP (high hydrostatic pressure) processing for food technology began when Hite (1899) demonstrated that the shelf life of milk and other food products could be extended by pressure treatment.

This is the technology of applying high hydrostatic pressure to materials by compressing the surrounding water and transmitting pressure throughout the product uniformly and rapidly (Hyashi, 1989). Use of high pressure in food processing is an extension of a technology that is commonly employed in many other industrial processes, notably in the manufacturing of ceramics, diamonds, super-alloys, simulators and sheet metal forming. Similarly, high isostatic pressures are routinely used in the manufacturing of polymeric compounds, such as for the synthesis of low-density polyethylene and in chemical reactors for the manufacturing of quartz crystals.

The advances achieved in ceramics and metallurgical industries in the use of HHP techniques during the 1970's and 1980's, have led to the possibility of treating food by this method at industrial level. The first commercial HHP treated products appeared on the market in 1991 in Japan, where HHP processing is now being used for products such as fruit juices, jams, sauces, rice, cakes, and desserts. There is increasing worldwide interest in the use of HHP because of the advantages of this technology over other methods of processing and preservation. An important issue in the application of high pressure technology in the food preservation/processing industry is regulatory approval, which focuses upon microbiological and toxicological safety of food products. Generally high pressure preservation processes reduce the microbial load to the same level achieved by traditional technologies, while delivering higher-quality products. Combination of high pressure with other treatments (e.g., mild temperature elevation, refrigerated storage, and acidification) during processing and storage is a likely route the food industry will take, as the inactivation of bacterial spores and some pressure resistant enzymes at room temperature cannot be achieved by pressure alone. In high pressure processing of foods, foods are subjected to ultra high hydrostatic pressure (UHHP or UHP), generally in the range of 100-1000 MPa. The processing temperature during pressure treatments can be adjusted from below 0 °C to above 100 °C with exposure times ranging from a few seconds to over 20 min. The UHP process is usually carried out with water as a hydraulic fluid to facilitate the operation and compatibility with food materials (Earnshaw, 1996). Pressure primarily reduces the volume of a system.

Under equilibrium conditions, according to the Le Chatelier principle, processes associated with volume decrease are encouraged by pressure, whereas processes involving volume increase are inhibited by pressure (Butz and Tauscher, 1998). Due to this fact, at a relatively low temperature (0 - 40 °C) covalent bonds are almost unaffected by HP where the tertiary and quaternary structures of molecules which are maintained chiefly by hydrophobic and ionic interactions are altered by high pressure >200 MPa (Hendrick et al., 1998).

High pressure can be used in food processing in a similar way as temperature. For instance, hydrostatic pressure can induce gel formation in egg white and yolk, crude carp actomyosin, and a suspension of soy protein by the application of 1000–7000 atm pressure at 25 °C for 30 min (Farr, 1990). The most unique property of HHP is its ability to be transferred instantly and uniformly throughout food system. Thus, the application of HHP is independent of sample mass and geometry. Other important advantages in using this technology in food industry are:

- *Inactivation of microorganisms and enzymes*
- *Modification of biopolymers*
- *Quality retention, such as colour and flavour*
- *Changes in product functionality (Knorr, 1993).*

In order to implement this new technology in the food industry, we need to understand the mechanism and kinetics of pressure—induced degradation / denaturation / inactivation of several food compounds (e.g. nutrients, proteins, microorganisms, enzymes) and the way in which the degradation / denaturation / inactivation is induced by other parameters (e.g. temperature, pH).

Principle of High Hydrostatic Pressure

High hydrostatic pressure technology is based on the use of pressure to compress food located inside a pressure vessel. The pressure vessel is the most important component of HHP equipment, consisting of a forged monolithic cylindrical piece built of alloy steel with high tensile strength. Multilayer or wire-wound prestressed vessels are used for pressures higher than 600 MPa.

Figure 1.1 ***Commercial-Scale, High-Pressure Processing System***
(Horizontal machine)
(Source: Photo courtesy of NC Hyperbaric)

Prestressed vessels are purposely designed with residual compressive stress in order to lower the maximum stress level in the vessel wall during pressurization, hence reducing the cost of producing this important piece of equipment. In HHP equipment utilized in food applications, pressure is transmitted by two methods: direct or indirect. In the direct method, a piston is pushed at its larger diameter end by a low-pressure pump, directly pressurizing the pressure medium at its smaller diameter end. This method allows very fast compression but requires a pressure-resistant dynamic seal between the piston and the internal vessel surface to avoid leaks and contamination of the food. In the indirect method, high-pressure intensifiers are used to pump the pressure medium from the reservoir into the closed vessel until the desired pressure is achieved.

The applied pressure is isostatically transmitted by a fluid. In this way, uniform pressure from all directions compresses the food, which then returns to its original shape when the pressure is released.

Various physical and chemical changes result from the application of pressure. Generally, physical compression during pressure treatment results in a volume reduction and an increase in temperature and energy (Heremans, 2003). In predicting the effect of HPP on foods, it is necessary to consider the net combined pressure-temperature effect of the process. The following principles govern the behaviour of foods under pressure:

- **Le Chatelier's principle** *Any phenomenon (phase transition, change in molecular configuration, chemical reaction) accompanied by a decrease in volume is enhanced by pressure. Accordingly, pressure shifts the system to that of lowest volume (Farkas and Hoover, 2000).*

- **Principle of microscopic ordering** *At constant temperature, an increase in pressure increases the degree of ordering of molecules of a given substance. Therefore, pressure and temperature exert antagonistic forces on molecular structure and chemical reactions.*

- **Isostatic principle** *Pressure is uniformly distributed throughout the entire sample, whether in direct contact with the pressurizing* *medium or insulated from it in a flexible container. Thus, the process time is independent of sample size and shape, assuming uniform thermal distribution within the sample.*

If a food product contains sufficient moisture, pressure will not damage the product at the macroscopic level as long as the pressure is applied uniformly in all directions. For example, a grape can be easily crushed if pressure is applied to it by placing it between two fingers and squeezing along one axis. In contrast, if the grape is exposed to a uniform pressure by submerging it in water inside a sealed flexible bottle and squeezing, the grape retains its shape no matter how hard the bottle is squeezed. In this case, the pressure transmitted from the bottle wall through the water is applied uniformly around the fruit. Similarly, pressure will not damage most foods processed by high pressure, providing the food does not have a porous structure containing air voids. Air pockets will collapse under pressure due to differences between the compressibility of air and water, and unless the food is perfectly elastic and consists of closed-cell foam from which air cannot escape, the food will not be restored to its original size and shape. As a result, foods like strawberries are crushed by HPP, but an air-filled balloon does not burst.

High Pressure Equipment

Although the principles of high-pressure processing for microbial inactivation have been known since the late 1800's (Hite, 1899), it is only relatively recent developments in mechanical engineering that have permitted large high-pressure vessels to be constructed at reasonable cost with sufficient durability to withstand thousands of pressure cycles without failure. High-pressure processing systems were initially developed in the chemical and material process industries for applications such as making artificial diamonds and sintered materials from powders. It is only during the past two decades that the food industry has begun using pressure treatment for food preservation. HPP is primarily practiced as a batch process where pre-packaged food products are treated in a chamber surrounded by water or another pressure-transmitting fluid. Semi-continuous systems have been developed for pump-able foods where the product is compressed without a container and subsequently packaged "clean" or aseptically.

The primary components of an HPP system include a pressure vessel; closure(s) for sealing the vessel; a device for holding the closure(s) in place while the vessel is under pressure (e. g., yoke); high-pressure intensifier pump(s); a system for controlling and monitoring the pressure and (optionally) temperature; and a product-handling system for transferring product to and from the pressure vessel. Normally, perforated baskets are used to insert and remove pre-packaged food products from the pressure vessels. Systems also have provisions for filtering and reusing the compression fluid (usually water or a food-grade solution). Commercial batch vessels have internal volumes ranging from 30 to more than 600 liters.

Avure Technologies, NC Hyperbaric, and Uhde are major suppliers of commercial scale pressure equipment. Both horizontal and vertical pressure vessel configurations are available. Commercial-scale, high pressure processing systems cost approximately $500,000 to $2.5 million, depending on equipment capacity and extent of automation. Currently, HPP treatment costs are quoted as ranging from 4–10 cents/lb, including operating cost and depreciation, and are not "orders of magnitude" higher than thermal processing—as is often thought (Sàiz et al., 2008). As demand for HPP equipment grows, innovation is expected to further reduce capital and operating costs.

A typical HPP process uses food products packaged in a high-barrier, flexible pouch or a plastic container. The packages are loaded into the high-pressure chamber. The vessel is sealed and the vessel is filled with a pressure transmitting fluid (normally water) and pressurized by the use of a high-pressure pump, which injects additional quantities of fluid. The packages of food, surrounded by the pressure-transmitting fluid, are subjected to the same pressure as exists in the vessel itself. After holding the product for the desired time at the target pressure, the vessel is decompressed by releasing the pressure-transmitting fluid (Farkas and Hoover, 2000). For most applications, products are held for 3–5 min at 600 MPa. Approximately 5–6 cycles/hr are possible, allowing time for compression, holding, decompression, loading, and unloading. Slightly higher cycle rates may be possible using fully automated loading and unloading systems.

After pressure treatment, the processed product is removed from the vessel and stored/distributed in a conventional manner. Liquid foods can be processed in a batch or semi-continuous mode. In the batch mode, the liquid product is pre-packaged and pressure-treated as described above for packaged foods. Semi continuous operation requires two or more pressure vessels, each equipped with a free-floating piston that allows each vessel to be divided into two chambers. One chamber is used for the liquid food; the other for the pressure-transmitting fluid. The basic operation involves filling one chamber with the liquid food to be treated. The fill valve is closed and then pressure-transmitting fluid is pumped into the second chamber of the vessel on the opposite side of the floating piston. Pressurization of the fluid in this second chamber results in compression of the liquid food in the first.

After an appropriate holding time, the pressure is released from the second chamber. The product discharge valve is opened to discharge the contents of the first chamber, and a low-pressure pump injects pressure-transmitting fluid into the second chamber, which pushes on the piston and expels the contents of the product chamber through the discharge valve. The treated liquid food is directed to a sterile tank from which sterile containers can be filled aseptically. Typically, three pressure vessels are used to create a semi-continuous system capable of delivering a continuous product output. This is accomplished by operating the three vessels such that one is loading, one is compressing, and one is discharging at any point in time (Farkas and Hoover, 2000).

The US food industry has gained about 18 years of experience with commercial scale, high-pressure equipment since the introduction of high-pressure pasteurization systems during the 1990's. In response to industrial demand for higher volume and throughput, high-pressure equipment vendors have focused on improving equipment productivity. Developments have included a doubling of pressure vessel volumes, automated systems for package and vessel loading, and higher horsepower intensifiers to speed up the compression process. Tandem machines (Sàiz et al., 2008), where multiple vessels are serviced by a common bank of high-pressure pumps, have also become available.

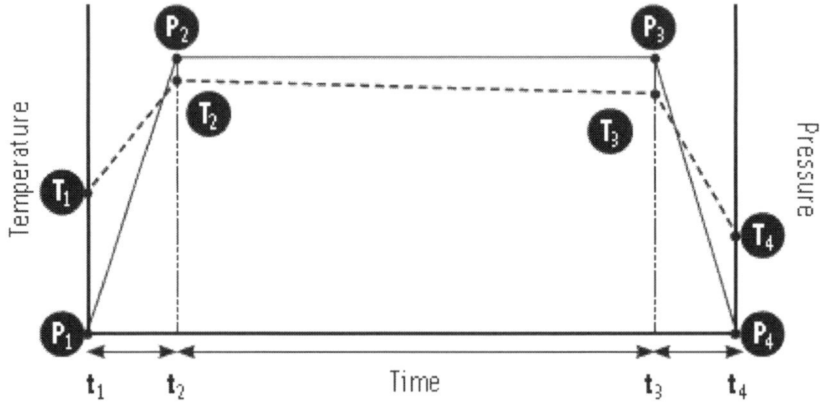

Figure 1.2 ***Sample Pressure-Temperature History During High Pressure Processing***
(Courtesy: V. M. (Bala) Balasubramaniam, Daniel Farkas, & Evan J. Turek)

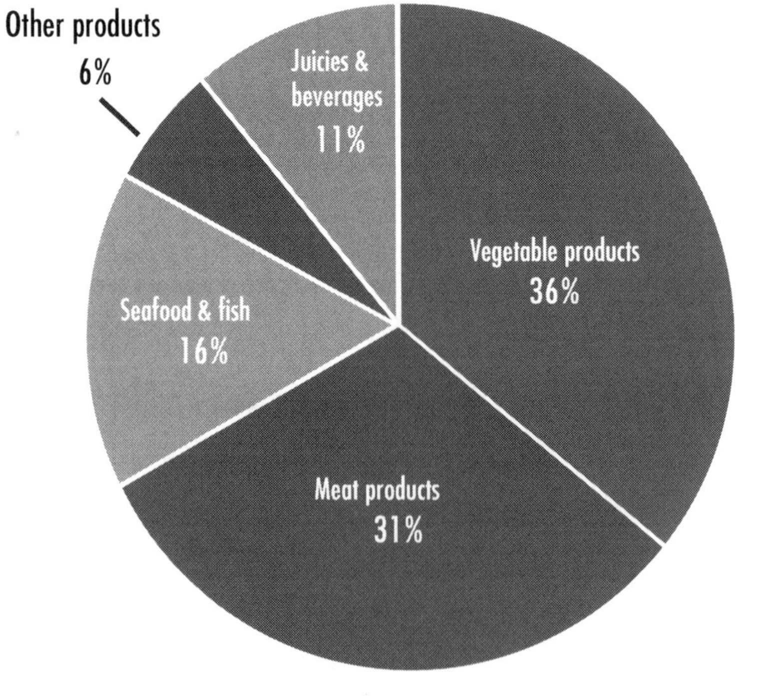

Figure 1.3 ***Percentage of Industrial High-Pressure Process Equipment Dedicated to Various Food Manufacturing Segments*** *From Samson (2008). (Courtesy: V. M. (Bala) Balasubramaniam, Daniel Farkas, & Evan J. Turek)*

By sequencing the compression cycles of multiple vessels, it is possible to increase throughput and reduce capital costs by shortening the compression time (through use of multiple pumps) and utilizing the same pumps to pressurize the other vessels during a period of time when the pumps would normally be idle in a single vessel system. With two or more pressure vessels operating under typical food processing conditions, a throughput of approximately 20 million lb/yr is achievable.

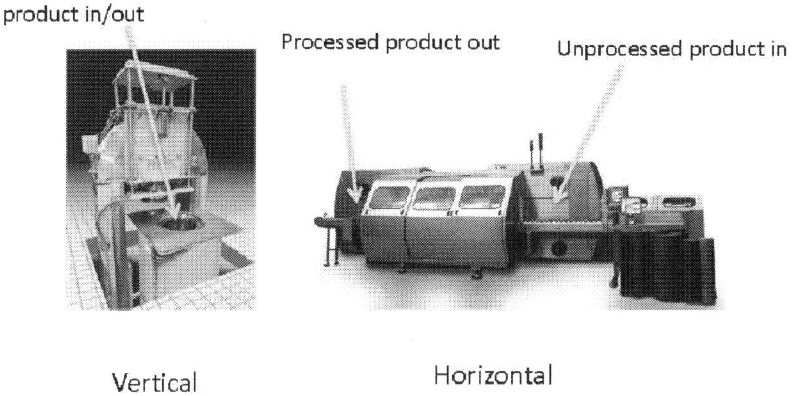

Figure 1.4 ***HPP Machines***
(Source: University of Queensland, Australia; Rnajan Sharma, 2011)

High Pressure Equipment Designs

Although the effects of isostatic pressure on microorganism inactivation and on protein and polysaccharide denaturation have been known over 100 years ago since Hite, because of technical difficulties and costs associated with high pressure processing units and packaging of materials, the food processing industry did not become interested in applying this technology until recent decades. The interest on the high pressure processing / preservation of foods has been triggered by results of research at universities and institutes and by the efforts of equipment suppliers who recognize a promising new market in high pressure technology. Although high pressure technology is currently more expensive than traditional processing technologies (e.g., high-temperature sterilization), the use of high pressure offers new opportunities for food industry to respond to consumers wishes (Hendrickx et al., 2005a).

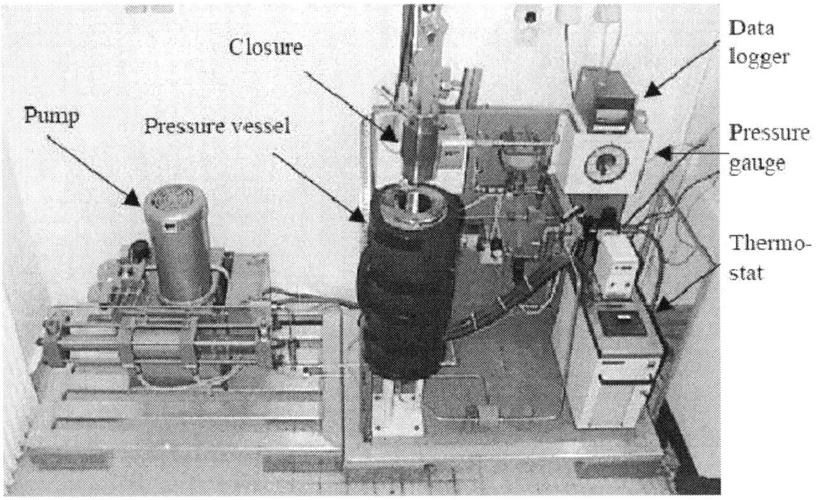

Figure 1.5 ***A Lab-Scale High Hydrostatic Pressure System***
The maximum pressure of the system is 600 MPa with a filling volume of 0.6 L, Temp.: 0 °C to 95 °C.
(Source: National Forge Europe, Belgium). Redrawn from Dong-Un Lee, 2002)

Generation and Delivery of High Hydrostatic Pressure

Pressure is a thermodynamic variable present in the biosphere. In nature, hydrostatic pressure increases by 1 atm pressure for every 10 m depth of water. Thus the pressure at the bottom of Mariana Trench, one of the deepest seas in our planet, reaches up to 116 MPa (Yayanos, 1998). However, HHP in food processing uses the pressure range of 100 - 1000 MPa, which is almost the same level or even ten times higher than the pressure of the deepest sea. Thus, special equipment is needed to generate and endure such high pressures.

A typical HHP system consists of a high pressure vessel and its closure, a pressure generation system, a temperature control device (Figure 1.5). The heart of the HHP system is obviously the pressure vessel, which is in many cases, is a forged monolithic, cylindrical vessel constructed in low alloy steel of high tensile strength. The wall thickness of the monobloc vessel determines the maximum working pressure. Depending on the internal diameter of the vessel, the use of monobloc vessels is typically limited to maximum working pressure of 400-600 MPa. In case higher pressures are required, pre-stressed vessel designs like multilayer vessels or wire-wound vessels are used. HHP can be generated either by direct compression (Figure 1.4) and indirect compression (Figure 1.5).

In the case of direct, piston-type compression, the pressure medium in the high pressure vessel is directly pressurized by a piston, driven at its larger diameter end by a low pressure pump. The indirect compression method uses a high pressure intensifier which pumps the pressure medium from the reservoir into the closed and de-aerated high pressure vessel, until the desired pressure is reached. Most of the industrial cold, warm and hot isostatic pressing systems use the indirect pressurization method.

The cold isostatic pressing (CIP) process can be either "wet bag (free mould)" in which the mould is filled outside the pressure vessel and then placed into the pressure vessel, and directly immersed in the fluid. These installations are used for pressing larger parts and complex forms. Or "dry bag (fixed mould)" in which case, the mould is fixed inside the pressure vessel and filled in situ. It is separated from the pressure fluid by an elastomeric sleeve. These machines are mostly used for smaller parts such as tubes, rods, and nozzles. For food processing applications, both dry-bag and wet-bag processes are of interest but wet bag is more suitable for food processing. A variety of flexible packaging may be used to contain samples for high-pressure processing.

Common packaging materials used in food high pressure processing, are ethylene vinyl alcohol copolymer (EVOH) and polyvinyl alcohol (PVOH). Pressure transmitting fluids are used in the vessel to transmit pressure uniformly and instantaneously to the food sample. Some of the commonly used pressures transmitting fluids are water, food-grade glycol–water solutions, silicone oil, sodium benzoate solutions, ethanol solutions, inert gases, and castor oil. The sealed packages are sufficiently flexible to withstand the compression which occurs during pressurization. The pressure medium and the pack contents are compressed to about 80–90% of their original volumes during pressurization in the 400–800 MPa pressure range, but of course, return to their original volumes when the pressure is released. Also, because of the low volume contraction of liquids and solids, particulate products such as foods are not mechanically damaged during pressurisation. The elastic capacity of many foods helps them to recover their original structure and shape (Rovere, 2001).

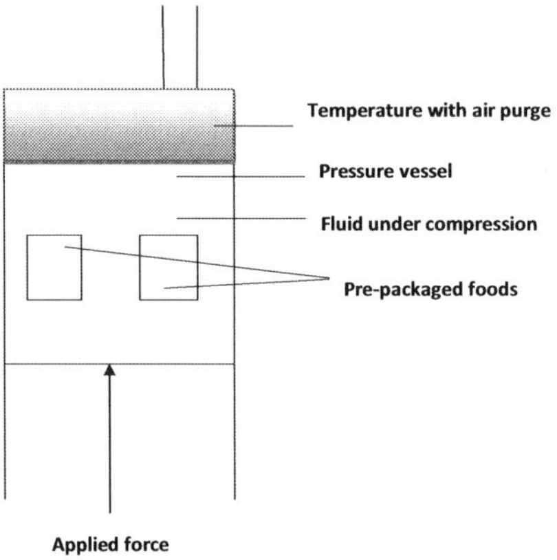

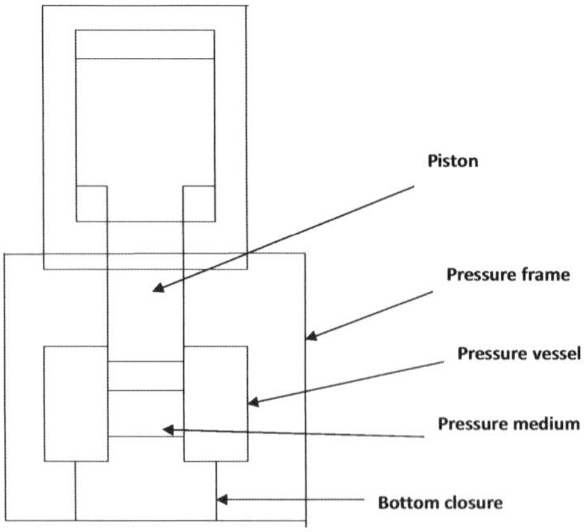

Figure 1.6 ***Direct Method for Generation of High Isocratic Pressure***
Uses a piston driven by a low-pressure pump.
(Source: African Journal of Biotechnology Vol. 7 (16), pp. 2739-2767, 18)

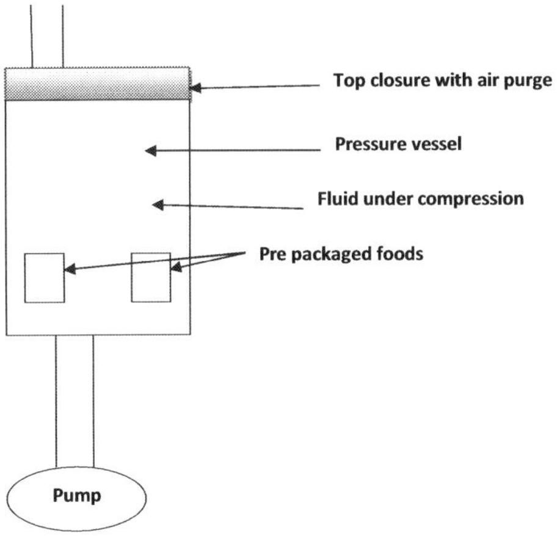

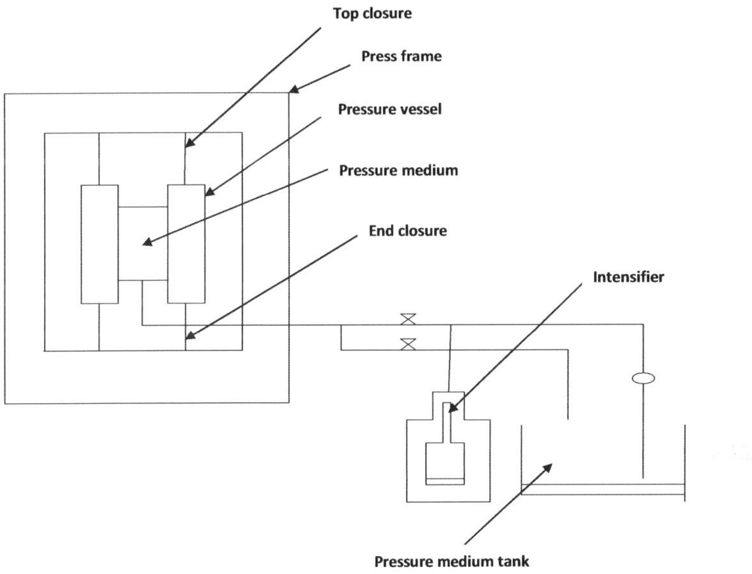

Figure 1.7 ***Indirect Method for Generation of High Isostatic Pressure***
Uses a high-pressure intensifier to pump pressure medium into closed vessel.
(Source: African Journal of Biotechnology Vol. 7 (16), pp. 2739-2767, 18)

There are three major types of high pressure processing of food products: the (conventional) batch systems derived from cold isostatic processing, semi continuous and continue systems. Batch systems can process both liquid and solid products, but these have to be pre-packed. In-line systems can be applied only to pumpable products (e.g. fruit juice). The product is pumped into the pressure vessel and pressurised using a floating piston, which separates the product from the pressure medium. For batch systems, the overall cycle time is the sum of the number of single steps: filling, closing, pressure built up, pressure holding, pressure releasing, opening, taking out (Figure 1.4)

For liquid products continuous treatment also is possible using tube reactors or special valve systems (Van den Berg et al., 2002). In a packed product immersed in a compressed liquid, the pressure is transferred homogeneously and instantaneously throughout the product.

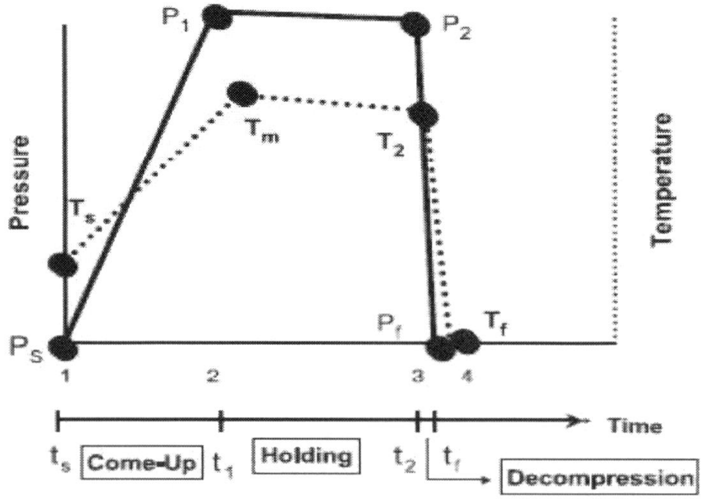

Figure 1. 8 ***Typical Variables (Pressure, Temperature, and Time) to Indicate Conditions of HPP Testing***

Ambient pressures Ps and Pf are typically 0.1 MPa. Tm is the maximum temperature at process pressure. The difference between ambient temperatures before and after high pressure processing. (Ts and Tf) generally indicates the extent of heat loss during testing (with the assumption that the depressurization time is within a few seconds). *Redrawn from Balasubramaniam et al 2004.*

(Source: African Journal of Biotechnology Vol. 7 (16), pp. 2739-2767, 18)

Under these conditions, the product is iso-pressed, since its internal pressure becomes equalised to the external one. In practical terms, the gaseous part of the product nearly disappears completely while the liquids and solids remain according to their compressibility.

Processing or Formulation Factors That Can Modify the Effect of High Pressure

Temperature

- *High temperatures increase the effect of high pressure against microorganisms and meat components.*

- *Sub-zero temperatures can protect fish products components, keeping reasonable microbial inactivation capacity.*

pH

- *Low pH values increase the effect of high pressure against micro-organisms and meat components.*

Bacteriocins

- *Some bacteriocins, and specially nisin, are very effective combined with high pressure, even on Gram negative bacteria.*

Water activity

- *With higher water activity, more effectiveness of high pressure processing, but also more recovery from sub lethally injured microorganisms.*

Preservatives

- *Lactate addition reduce the initial inactivation of the microorganisms but it delays its later recovery.*

Combination with other technologies

- *Combined processes can improve high pressure effects on microorganisms.*

Present Range of Pressures Available

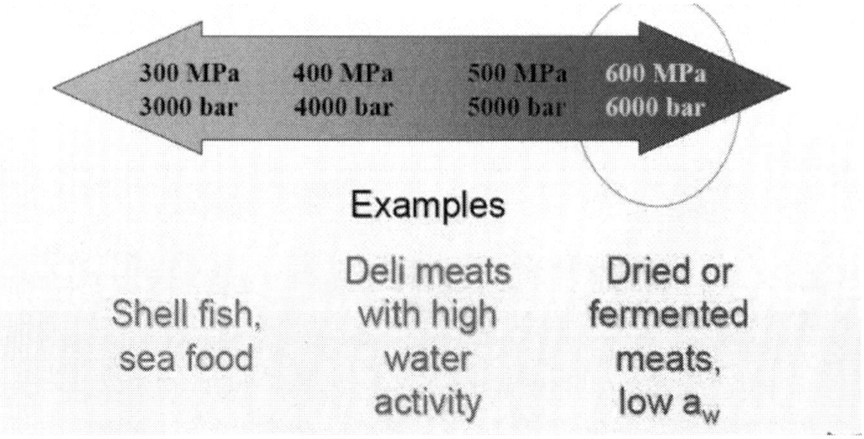

Processing Temperatures

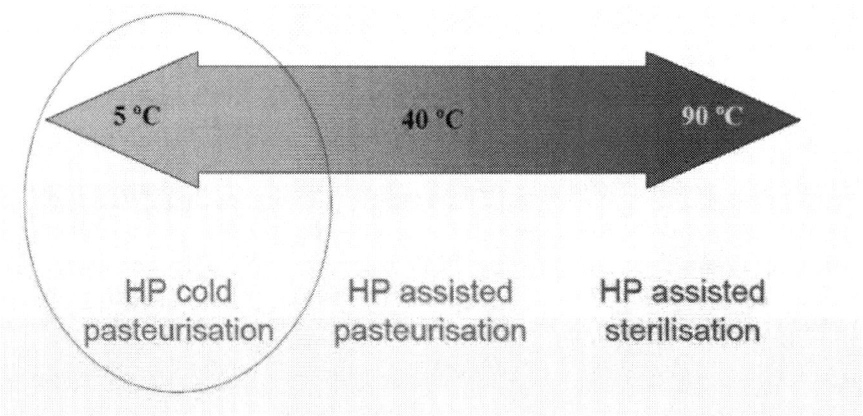

Food That Can Be HHP Treated

Solid foods, mainly vacuum packed

- *Dry-cured or cooked meat products*
- *Cheeses*
- *Fish, seafood, marinated products*
- *Ready to eat meals, sauces*
- *Fruits, marmalades / jams*
- *Vegetables*

Liquid foods, in flexible packaging

- *Dairy products*
- *Fruit juices*

Main Technological Effects of HHP in Meat Products

About colour

- ***In fresh or marinated meat, the iron in the myoglobin changes from ferrous to ferric and globin is denatured: the red colour is lost.***

About texture

- ***Inhibition or stimulation of the proteolytic activity in muscles activity muscles (depending on processing conditions).***

- ***Proteins are partially denaturized in products where proteins have not been previously modified by other process: heating, drying, and fermentation.***

Figure 1.9 ***Top View of the Fresh and Frozen Beef Samples Treated by HHP***
(Source: IRTA)

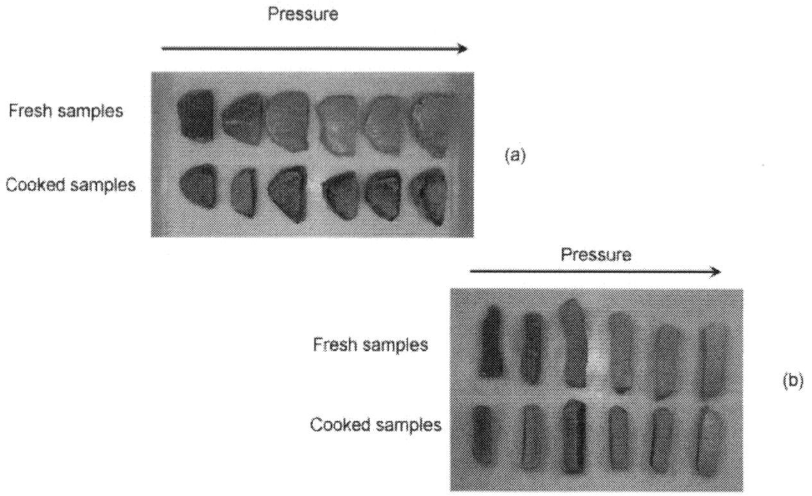

Figure 1.10 ***Cooked Samples: View From the Top (a); View of the Inside (b)***
(Source: IRTA)

Figure 1.11 ***Commercial Beef Products Before and After HHP Treatment***
(Source IRTA)

Examples of Uses of HHP

Figure 1.12 ***Cured Meat Products***
(Source IRTA)

Figure 1.13 ***Cooked Meats: Cooked Ham, Roasted Chicken, Turkey Breast, Cooked Causages***
(Source IRTA)

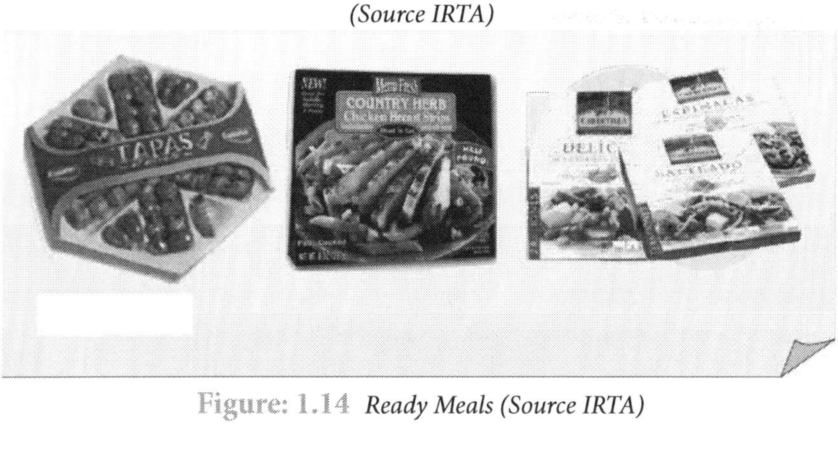

Figure: 1.14 *Ready Meals (Source IRTA)*

Advantages and Limitations of High Pressure Processing

Some of advantages and limitations of using hydrostatic pressure method are:

Some advantages of HHP

- *High pressure is not dependent on size and shape of the food.*

- *High pressure is independent on time/mass, that is, it acts instantaneously thus reducing the processing time.*

- *It does not break covalent bonds; therefore, the development o flavours alien to the products is prevented, maintaining the natural flavour of the products.*

- *It can be applied at room temperature thus reducing the amount of thermal energy needed for food products during conventional processing.*

- *Since high pressure processing is isostatic (uniform throughout the food), the food is preserved evenly throughout without any particles escaping the treatment.*

- *The process is environment friendly since it requires only electric energy and there are no waste products.*

Some limitations of HHP

- *Food enzymes and bacterial spores are very resistant to pressure and require very high pressure for their inactivation.*

- *The residual enzyme activity and dissolved oxygen results in enzymatic and oxidative degradation of certain food components.*

- *Most of the pressure-processed foods need low temperature storage and distribution to retain their sensory and nutritional qualities.*

Effects of High Hydrostatic Pressure on Microorganisms

Microbial inactivation is one of the main tasks for the application of high pressure technology. Many reports have demonstrated the inactivation effect of HHP on microorganisms, extending in this way the microbial shelf life and improving the microbial safety of food products. It has been demonstrated that bacterial spores require more extreme conditions of pressure, temperature, and longer treatment times to be inactivated . In general, vegetative cells are inactivated at low pressure levels, around 400 - 600 MPa, while more resistant bacterial spores can survive pressures higher than 1000 MPa .

Inactivation Mechanism

High hydrostatic pressure brings about a number of changes in the morphology, cell membrane or biochemical reactions of microorganisms, and all these processes are related to the inactivation of microorganisms. For instance, Hamada et al. (1992) noticed changes in colony form after pressure treatment of *Saccharomyces cerevisiae.* Especially, the cell membrane is considered to be the major target for the pressure-induced inactivation of microorganism, and it is generally accepted that the leakage of intracellular constituents through the permeabilized cell membrane is the most direct reason of cell death by high pressure treatment. However, if applied pressure was not severe enough to induce a total permeabilization of cell, the permeabilization took place only in the outer membrane, in the case of gram negative bacteria the permeabilized membrane was rapidly restored after pressure release.

The fluidity of cell membrane plays an important role in the susceptibility of microorganisms to pressure treatments. Microorganisms with less fluid membranes are more sensitive to HHP Conversely, increased membrane fluidity protected against pressure inactivation. The denaturation of key enzymes in microorganism by pressure has been regarded by another important reason of cell death.

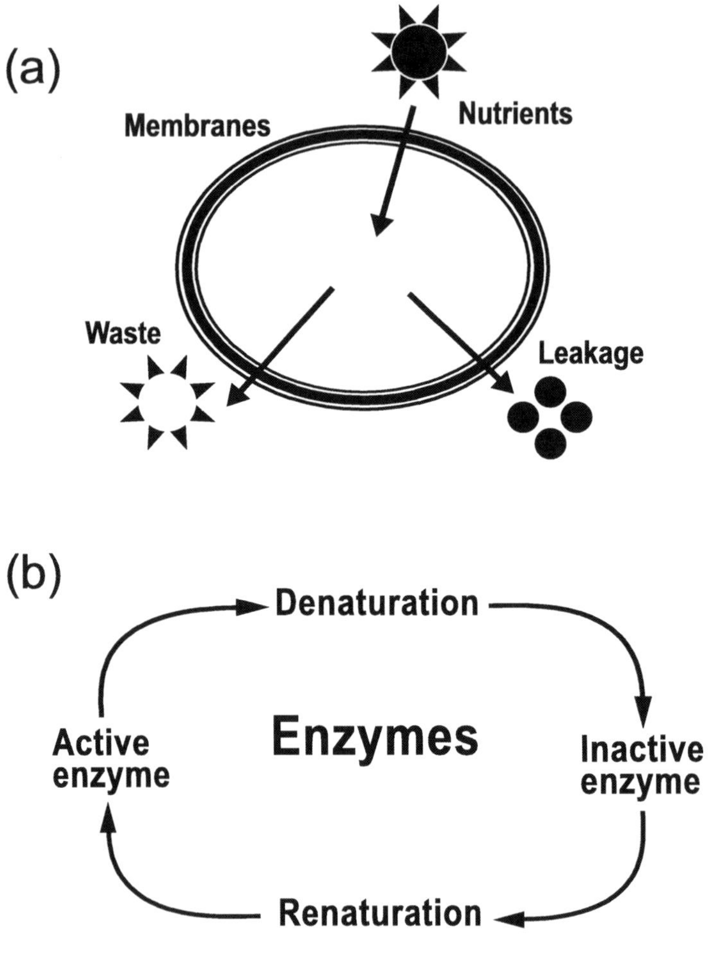

Figure 1.13 ***Hydrostatic Pressure Effects on Selected Microbial Functions***
(a) Cellular membranes; (b) Microbial enzymes
Redrawn from Torres, et al. (2005)
(Source: African Journal of Biotechnology Vol. 7 (16), pp. 2739-2767)

Because intracellular enzymes seemed not to be a determining factor of pressure resistance, membrane bounded ATPase was considered to such a key enzyme. Bacterial spores are known to be pressure resistant, and the inactivation mechanism is different from that of other vegetative microorganisms. It was assumed that pressure caused inactivation of spores by first initiating germination and then inactivating germinated forms. Hydrostatic pressures of 100–300 MPa can induce spore germination and resultant vegetative cells are more (Yaldagard et al., p. 2745) sensitive to environmental conditions. Biochemically, binding of germinant to its receptor is believed to promote the germination process followed by the efflux of Ca^{+2} and other ions and the influx of water into the spore, which result in the activation of pore specific cortexlytic enzyme (Wuytack et al., 2000).

This germination process can be enhanced by pressure treatments (as a general rule (Le Chateler's principle)), because the volume of system decreases during germination as a result of increased solvation of spore's components. It was also suggested that pressure-induced germination involves activation of physiological pathway and is therefore not merely a physicochemical process in which water is forced into the spore protoplast, because inhibitors of nutrient-induced germination also inhibit pressure induced germination (Wuytack et al., 2000).

However, when high pressure is applied to spores with elevated temperature, the cortexlytic enzymes in spores could be directly inactivated and the inactivation of spores took place without the germination step (Heinz and Knorr, 2001). The overall pattern of spore inactivation showed a strong pressure–heat synergism. The effect has been confirmed for a wide range of spore types, although the effectiveness of the combination varies greatly in magnitude for different spores.

The kinetics of pressure inactivation was approximately exponential for spores of *Bacillus pumilus* (Clouston and Wills, 1970), but for spores of *Bacillus coagulans, Bacillus subtilis* and *Clostridium sporogenes* concave upward curves or long tails were reported. The fact that, although spores of some species are relatively sensitive to pressure (e.g. Bacillus cereus), those of other species, including some of special importance in foods such as *Bacillus stearothermophilus* and *Clostridium botulinum*, are very resistant; has so far prevented the use of pressure to sterilize foods. This situation may change with the development of presses that operate at higher temperature–pressure combinations, or the development of other effective combination techniques. For instance, the presence of bacteriocins such as nisin can amplify the effect of pressure against spores, for example, of *B. coagulans.*

Critical Parameters for Microbial Inactivation by High Hydrostatic Pressure

Primary Factors

Besides type of microorganism, composition of suspension media or food, pressure level and treatment time, the critical parameters for high pressure induced microbial inactivation are pH, water activity (aw) and the treatment temperature. Various combinations of these parameters have been investigated and the general rules are as follows:

- *Microorganisms become more susceptible to pressure at lower pH and recovery of subletally injured cells is reduced. Further, sub lethally injured microorganisms induced by HHP can be reactivated in a nutrition-rich environment, but fail to repair at acidic conditions.*

- *A reduction of water activity exerts protective effect for microorganisms against pressure treatments.*

- *The treatment temperatures above or below room temperature tend to increase the inactivation rate of microorganisms.*

Secondary Factors

Other factors influence the effectiveness of HPP. For example, the redox potential of the pressure menstruum may also play a role in the inactivation for some microorganisms (Yaldagard et al., p. 2747).

In addition to the above, a factor that significantly influences the effectiveness of HHP treatment on the inactivation and consequently the reduction in microbial population is the medium composition in which microorganisms are dispersed. Food constituents such as sucrose, fructose, glucose and salts affect the baroresistance of microorganisms present in food.

Effects of High Pressure on Chemical Reactions Related to Food Quality

The most basic concept to interpret the effects of pressure on chemical reactions is the Principle of Le Chatelier. Pressure primarily reduces the volume of a system. Under equilibrium conditions, according to the Le Chatelier principle, processes associated with volume decrease are encouraged by pressure, whereas processes involving volume increase are inhibited by pressure.

Pressure as an important thermodynamic variable can affect a wide range of biological structures and processes. Since the mechanism of UHP is based on decrease in volume, high pressure enhances the rates of chemical and biochemical reactions. Moreover, reactions involving formation of hydrogen bonds are favoured by high pressure because bonding results in a decrease in volume of the molecules. UHP does not break down the covalent hydrogen, ionic or hydrophobic bonds. Covalent bonds are resistant to pressure, which means that low molecular weight food components responsible for nutritional and sensory characteristics remain intact during pressure treatment, whereas high molecular weight components whose tertiary structure is important for functionality determination are sensitive to pressure.

With respect to food systems important characteristics of high-quality foods are texture, flavour, colour, and nutritive value. The first three properties are correlated with purchase and consumption quality and determine consumer's acceptance of the product. Nutritive value (i.e. vitamins, minerals, and other nutrients) is a hidden quality.

Chemical or biochemical reactions occurring in food products can bring about undesirable changes in or deterioration of these attributes of food quality during preservation / processing treatments and subsequent storage. High pressure allows inactivation of pathogenic / spoilage microorganisms and food-spoiling enzymes while leaving most attributes of food quality intact This advantage is attributed to the fact that high pressure keeps covalent bonds intact and affects only non covalent bonds (such as hydrogen, ionic, and hydrophobic bonds).

The Effect of High Pressure on Texture, Flavour, Colour, and Nutritive Values of Food Products

The colour retention effects of high pressure for many fruit and vegetable products such as orange juice, fruit jam, tomato juice investigated by some researchers. Chemical and spectrophotometrical analysis showed that high pressure treatment largely preserves fresh colour. Also, the effect of high pressure on meat and meat products has been studied by many authors because high pressure processing may offer potential to preserve and restructure meat, provided that, the red colour can be maintained. For fruit jams (e.g., strawberry jam) high pressure was found to retain fresh flavour much more than traditional thermal processing (Dervisi et al., 2001).

In the experiments carried out by Zabetakis et al. (2000) on the effects of high hydrostatic pressure on strawberry flavour compounds, the highest flavour stability was observed when samples were treated with lower pressures <800 MPa and they were stored at 4 and 30 °C. In studies conducted by Rodrigo et al. (2006), no colour degradation of tomato appeared under combined thermal and high pressure treatment (300–700 MPa, 60 min, 65 °C).

An extensive study on the effect of pressure on the texture of fruits and vegetables has been carried out by Basak and Ramaswamy (1998). They observed that, the change in firmness of treated samples to be dependent on both pressure level and pressurization time, (Suzuki et al.) studied the effect of pressure on water-soluble compounds responsible for taste and meaty flavour. The amount of peptides and amino acids estimated as phenol-reagent-positive materials apparently increase with increasing pressure up to 300 MPa (5 min at 2 °C).

The effect of pressure on the flavour of milk and dairy products has only been studied to a very limited extent. High pressure processes at 400–500 MPa for 3–15 min followed by refrigeration were found to result in a shelf-life comparable to that of thermal pasteurization; however, this report made no mention of sensory properties of the pressure-processed milk High pressure treatment only slightly influenced milk viscosity. The building up of milk viscosity can be described by a first-order reaction. Increase in viscosity has been explained to be the result of casein micelle disintegration. The properties of acid-set gels prepared from HHP treated milk have been reported by Johnston et al. Results indicate improved texture (rigidity and resistance to breaking) and syneresis resistance of the gels, measured by drainage or by centrifugation.

In studies conducted by Ancos et al. (2000), the effects of high pressure on the physicochemical, chemical, microbiological, and sensory characteristics of stirred low-fat yoghurt; laboratory-made yoghurts were treated at high pressure (100-400 MPa) for 15 min at 20 °C. Pressurized yoghurts exhibited higher viscosity and amino acid contents than did the untreated controls, and the differences were maintained after chilled storage.

The application of HHP technology to cheese milk causes differences in cheese composition and ripening in comparison to pasteurised milk cheese. The HHP-treated milk cheeses have higher moisture, salt, and total free amino acids contents than raw or pasteurised milk cheeses In relation to cheese texture and microstructure, Buffa et al. (2001), using uniaxial compression and stress relaxation tests, and confocal laser scanning microscopy showed that cheeses made from raw or HHP treated milk were firmer and less fracturable than cheeses made from pasteurised milk, but differences became less notable toward the end of ripening. Cheeses from pasteurised and HHP-treated milk were less cohesive than from raw milk. Although cheese exhibited a loss of elastic characteristics with ageing, cheeses from HP treated milk were the most elastic initially.

Effects of High Hydrostatic Pressure on Protein Systems

Proteins are linear polymers of amino acids, for which three or four levels of structural organization can be discerned. The primary structure, which is defined as the amino acid sequence and the location of disulphide bonds (if any), gives a complete description of the covalent bonds of a protein. The secondary structure refers to the way in which the polypeptide chain forms α helices or β sheets by intra- or intermolecular hydrogen bonds. The tertiary structure describes how the secondary structure domains fold into a three-dimensional configuration as a consequence of non-covalent interactions between amino acid side chains. The quaternary structure refers to the spatial arrangement of subunits, held together by non-covalent bonds between the polypeptide subunits (multi-meric proteins).

Proteins are delicate structures, maintained by interactions within the protein chain (determined by the amino acid sequence) and by interactions with the surrounding solvent. Changes in environmental conditions such as solvent composition, temperature, and pressure can perturb the subtle balance of intermolecular and solvent-protein interactions and can therefore, lead to (complete) unfolding / denaturation of the polypeptide chain. The term "denaturation of protein" indicates the phenomena in which the higher structure of protein is ruptured by environmental changes, while the primary structure is kept without damage.

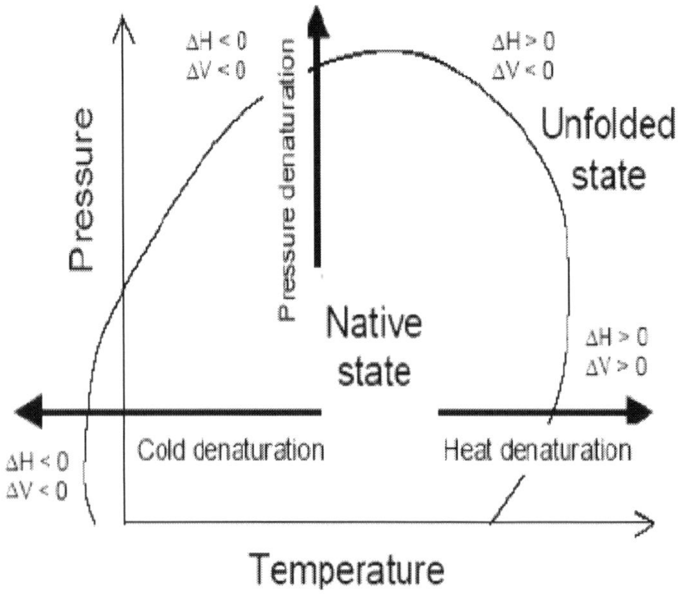

Figure 1.14 ***Typical Phase Transition Curve of Proteins in the PT-Diagram***
The relation between heat, cold and pressure-denaturation of proteins is presented by the sign of enthalpy changes (ΔH) and volume changes (ΔV).
(Redrawn from Heremans, 2002)

The effect of high pressure on proteins has been the subject of many reviews. Pressure treatment can cause a variety of effects on proteins, inducing reversible or irreversible structural modifications leading to protein denaturation, aggregation or gel formation. See Figure 1.14.

Structural rearrangements taking place in the protein upon pressurization are governed by the principle of Le Chatelier, which states that "processes associated with a volume decrease are encouraged by pressure increases, whereas processes involving a volume increase are inhibited by pressure increases". Gelation by high pressure treatment is due to a decrease in the volume of the protein solution; in contrast, heat denaturation of protein is caused by the violent movement of molecules leading to the destruction of non-covalent bonds.

Covalent bonds are unaffected by high pressure, thus the primary structure of proteins remains intact during pressure treatment. Secondary structures of proteins change at very high pressures, which might be explained by the cleavage of hydrogen bonds, which are enhanced at low pressures. These changes depend on the nature and concentration of the protein, as well as on the applied pressure, temperature, treatment time, and characteristics of the surrounding environment, such as pH or ionic strength (Rovere, 2001).

Pressure Versus Temperature Effects

Quantitative evaluation of pressure-induced protein denaturation proposed p/T diagrams (Figure 1.14) which revealed elliptic contours indicating that proteins can yield heat and cold denaturation. One of the practical consequences of this phenomenon is the stabilization against heat denaturation by low pressures. Not only has this been observed in several proteins and enzymes, but it also applies to the effect of pressure on the heat gelation of starch (Yaldagard et al. p. 2749).

A fundamental difference between pressure and temperature induced protein denaturation is that no change in covalent bonding has been observed in the pressure induced protein denaturation. Further, the structure of pressure denatured proteins differs significantly from that of heat denatured proteins. The pressure denatured proteins are relatively compact and retain elements of secondary structure while the heat denatured proteins have the extended, nearly random coil configurations (Ghosh et al., 2001).

Effects of High Hydrostatic Pressure on Food Protein

It is well-known that pressure affects proteins and leads to protein denaturation, aggregation or gel formation. The first systematic observation about the denaturation of proteins by high pressure was made by Bridgman (1914) treating egg albumin. He observed that the appearance of the pressure induced coagulum is quite different from that induced by temperature and the ease of pressure induced coagulation increases at low temperatures.

However, more systematical studies on the pressure induced denaturation of protein have been conducted after 50 years later using ovalbumin, chymotrypsinogen (Hawley, 1971), or metmyoglobin. Ovalbumin, the main component of egg white, was denatured under high pressure, as confirmed by the decrease in its α helical content and DSC endothermic enthalpies. The results indicated that only conformational changes are induced by pressure treatments because no change in PAGE pattern was observed. Ovalbumin does not form a gel when pressurised for 30 min at ~400 MPa.

This relatively high-pressure stability may be due to the presence of four disulfide bonds and strong non-covalent interactions stabilising the three dimensional structure of the protein. Gelation does occur, however, at higher pressures.

Doi et al. studied the effects of pressure on the stability of network bonds in heat-induced ovalbumin gels. They placed a steel ball on top of a heat-set gel and then pressure-processed. It was found that at pressures 400 MPa, the steel ball penetrated deep into the gels, suggesting that primarily hydrophobic and electrostatic interactions were destabilised in the gel, resulting in the partial melting of the gel. In contrast, cold-set gelatine gels did not melt under pressure treatment at 600 MPa, which suggested that hydrogen bonds were not destabilised by pressure processing.

Other researchers demonstrated that ovalbumin gels produced

by high pressure processing were more elastic and softer than heat-induced gels although the gels tended to become harder and less adhesive with increasing pressure. Taste and flavour of pressure-induced gels were free of cooked flavour and there was no destruction in vitamins and amino acids. Also, in the study conducted on the effect of pressure at different temperature (10, 25, and 60 °C) and pH (7.6 and 8.8) levels on selected properties of egg white solutions, it is observed that pressure induced an increase in turbidity, surface hydrophobicity, exposed SH content, and susceptibility to enzymatic hydrolysis, while it resulted in a decrease in protein solubility, total SH content, denaturation enthalpy, and trypsin inhibitory activity.

Moreover, it is reported that the pressure-induced changes in the selected properties were dependent on the pressure, temperature applied, and the pH (Plancken et al., 2005). Effects of pressure on milk proteins have been extensively investigated. A lactoglobulin, the main whey protein component is sensitive to high-pressure treatment. Solution studies of native α-lactoglobulin have indicated that pressure treatment has a notable effect on the protein's conformational and aggregation properties, which are more extensive at higher concentrations.

It has been reported that the extended application of high pressure up to 800 MPa has no influence on the surface activity of Calcium-free bovine α-casein. This behaviour related to the little ordered secondary and no tertiary structure of Calcium free. However, in milk itself, where the casein is complexed with colloidal calcium phosphate, the quaternary structure of the casein micelle is substantially affected by high pressure. The gel strength of high-pressure-induced whey protein concentrate gels as a function of pH has been studied by various researchers.

Electron microscopy has revealed that heat-induced gels (0.1 MPa for 30 min at 80 °C) are stronger and possess a greater number of permanent cross-links between the polypeptide chains. In contrast, high-pressure treatment (400 MPa for 30 min at 20 °C) generates fragile gels with a more porous network and fewer intermolecular cross-links. The application of pressure treatments to cheese processing was investigated in order to accelerate ripening period (Messens et al., 2000; Johnston and Darcy, 2000).

The rheological parameters and meltability of immature Mozzarella cheese were converged to those of ripened cheese by pressure treatment of 200 MPa due to casein dissociation and internal moisture redistribution. The hardness of cheese increased again above 200 MPa due to the protein denaturation and aggregation.

In vegetable proteins such as soy, pea, wheat gluten, and broad bean that are being widely used in the food industry for the formulation of new food products, high-pressure processing is considered to be a gentler processing operation in comparison to thermal processing. Matsumoto and Hayashi and Okamoto et al. have found that a minimum pressure of 300 MPa for 10–30 min is required to induce gelation.

It has also been reported that high pressure produces softer gels with a significantly lower elastic modulus than those made by heat treatment. Wheat gluten is often used for its unique cohesive and viscoelastic functional properties, with major application to wheat flours low in gluten.

To generate novel textures and products, Apichartsrangkoon et al. subjected hydrated wheat gluten to pressure heat treatment in the range 200–800 MPa at temperatures 20–60 °C with holding times from 20–60 min. The gels formed were very different from heat-processed gluten since they presented a more marked elasticity with high values of module of elasticity, which was derived from a 'Young's modulus' that was 2–3 times larger than the usual shear modulus (control).

The effects of combined high-pressure in the range of 300–700 MPa, and heat treatment (90 °C, 15 min) on the textural properties of soya gels was studied. It is observed when the solutions were pressurised before heat treatment, all the proteins formed self-standing gels (Molina and Ledward, 2003). In studies conducted by Kato et al. (2000), rice grains (Oryza sativa L. Japonica var. Akitakomachi) immersed in distilled water exhibited solubilisation and subsequent release of rice allergenic proteins in the range 100–400 MPa. In this range of pressure, considerable amount of proteins, 0.2–0.5 mg/g of rice, was released with maximum amounts obtained in the pressure range 300–400 MPa.

High pressure treatment at different temperatures will induce different effects on meat texture since the weak linkages stabilising the secondary, tertiary, and quaternary structures of a protein res, and differently to heat and pressure. Bouton et al. found that a pressure of about 100 MPa applied for 2.5 min or longer to post-rigor muscle at 40–60 °C improved the tenderness of the meat and Beilken et al. reported that pressure treatment at 150 MPa during treatment at 40–80 °C prevents the development of the myofibrillar component of toughness, but has little or no effect on the connective tissue component of toughness, other than to raise the temperature at which heat treatment alone produces a decrease in this component.

Fernandez-Martin et al. found that when pork batters were subjected to combinations of two pressures (200 and 400 MPa) and five heat treatments (10–70 °C); pressurization may stabilize the proteins against subsequent thermal denaturation. Also, the effects of high pressure (800 MPa) applied at different temperatures (20–70 °C) for 20 min on beef post-rigor longissimus dorsi texture were studied. Texture profile analysis showed that, when heated at ambient pressure there was the expected increase in hardness with increasing temperature and when pressure was applied at room temperature there was again the expected increase in hardness with increasing pressure. Myosin was relatively easily unfolded by both pressure and temperature, and when pressure denatured; a new and modified structure was formed of low thermal stability (Ma and Ledward, 2004).

Most of the studies related to the application of high pressure to sea foods have been conducted on its effects on fish proteins. The breakdown of ATP-related products is usually catalysed by certain dephosphorylases inherent in fish muscles. Shoji and Saeki observed a decrease in IMP level in carp muscle treated at 350 and 500 MPa and subsequent storage at 5. This is due to protein denaturation and deactivation of enzymes (involved in the degradation of ATP and related compounds) during high-pressure treatment.

High pressure treatment of surimi has produced a new gel product with excellent flavour, lustre, density, and elasticity, quite different from those treated by heat (Yoshioka and Yamada, 2002). Furthermore, the pressure-induced gels retained the natural qualities (that is, colour and flavour) of the raw material without the formation of cooked colour and flavour Pressure-induced surimi gels from marine species like pollack, sardine, skipjack, tuna, and squid have been reported to be smoother, more elastic and sensory superior to those produced by heat (Venugopal et al., 2001).

Effect of High Pressure on Enzymes

Enzymes are a special class of protein with an active site, formed by the three-dimensional conformation of molecules Generally, enzymes characterized by two striking properties their enormous catalytic power and their specificity quite often; the rate of an enzyme-catalyzed reaction is 106 to 1020 times that of an unanalysed one. Enzymes are highly specific, both in the type of reaction catalyzed and the choice of substrate. In view of the specificity of enzymatic reactions, enzymes may be affected by pressure in several ways:

- *Pressurization at room temperature may bring about reversible or irreversible, partial or complete enzyme inactivation resulting from conformational changes in the protein structure. These changes depend on the type of enzyme, micro-environmental conditions, pressure, temperature, and processing time.*

- *Enzymatic reactions may be enhanced or inhibited by pressure, depending on the volume change (positive or negative) associated with the reaction. Pressure induced changes in the catalytic rate may be due to changes in the enzyme-substrate interaction, changes in the reaction mechanisms, or the effect of a particular rate limiting step on the overall catalytic rate.*

- *A macromolecular substrate (protein, starch) may become more sensitive to enzymatic action once it has been unfolded or gelatinized by pressurization.*

- *Provided that the cell membrane or the membrane of Yaldagard et al. 2751 intracellular organelles is altered, intracellular enzymes may be released from extra-cellular fluids or cell cytoplasm hereby facilitating enzyme-substrate interactions.*

Mechanisms of Pressure Inactivation of Enzymes

High pressure can result in both activation and inactivation of enzymes depending on the pressure level and conditions. Usually at low pressure, enzyme activity may be enhanced, and at higher pressure the activity is inhibited Relatively low pressures of 100–200 MPa have been shown to activate monomeric enzymes, whereas higher pressures generally induce enzyme inactivation.

Butz et al. and Gomes et al. explained that beside conformational changes, the de-compartmentalization caused by pressure results in enzyme activation. When the food material is intact, there is compartmentalization between the enzyme and substrates. Application of low pressure in food tissue damages membrane and enzyme leakage leads to enzyme-substrate contact. Sometimes the activity of enzymes is enhanced or curtailed due to pH changes resulting from releasing of some components in the environment by the induced pressure. The effect of pressure on chemical or biochemical systems is described by the thermodynamical parameter (ΔV) (Equation 1.1):

$$\left(\frac{\delta\Delta G}{\delta P}\right) = -RT\left(\frac{\delta \ln K}{\delta P}\right) = \Delta V \qquad \text{Eq. 1.1}$$

the change of partial molar volume between initial and final state at constant temperature. It is governed by the principle of Le Chatelier, which predicts that application of pressure shifts equilibrium to the state that occupies the smallest volume. Hence, pressure favours reactions accompanied by a volume decrease and vice versa.

Kinetics of Pressure Inactivation of Enzyme

In general, the decrease of enzyme activity (A) as a function of processing time (t) can be written as seen in Equation 1.2.

$$\left(\frac{dA}{dt}\right) = -kA^{n} \qquad \text{Eq. 1.2}$$

Often, enzyme inactivation can be described by a first order kinetic model. It is suggested that, in the case of apparent first-order kinetics, one of the disruption formation of different interactions, decomposition of amino acids and aggregation / dissociation reactions in the enzyme inactivation predominates over the others.

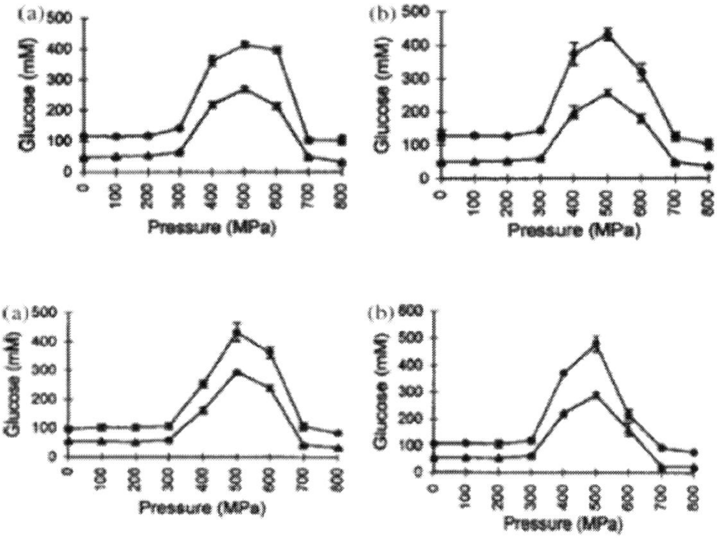

Figure 1.15 *Effect of HHP (100-800 MPa) for (a)10 min and (b) 20 min at Ambient Temperature on Amylase Activities in Wheat Four.*
(•) Total soluble carbohydrate
(Δ) Reducing sugars
(top) Barley flour
(bottom) Wheat flour
(Source: Gomes et al., 1998).

Several reactions occur more or less at the same rate, complex non first-order kinetics is expected as for pressure-temperature combination effects, a pressure-temperature kinetic data for some enzymes affecting texture, flavour or colour of food is exemplified in the following paragraph.

Reasons For the Industrial Use of HHP

- *Consumer satisfaction*
  - ~ *Fresher flavour*
  - ~ *Healthy or convenient attributes*
- *Legal requirements*
  - ~ *Related to food safety*
  - ~ *Related to export regulations*
- *As a reaction in front a food crisis*
- *To meet costumers' specifications*
- *Shelf life extension*
- *Image of innovative company*
  - ~ *Both to retailers and end users*
  - ~ *Producing safer and reliable products*
- *Technology fully accepted by consumers*

References

Raghupathy, Ramaswamy, Balasubramaniam, (Bala) V.M.; and Kaletunç, Gönül. *FSE-1-04, Extension Fact Sheet.* College of Food, Agricultural and Environmental Sciences, The Ohio State University, Columbus Publication. http://grad.fst.ohio-state.edu/hpp/. Web.

Yaldagard1, Maryam; Mortazavi, Seyed Ali; and Tabatabaie, Farideh. "The principles of ultra high pressure technology and its application in food processing / preservation: A review of microbiological and quality aspects." *African Journal of Biotechnology* Vol. 7 (16) :2739-2767. Print.

Balasubramaniam, (Bala) V.M.; Farkas, Daniel, & Turek, Evan J. *Preserving foods by destroying pathogenic and spoilage organisms while keeping food chemistry basically intact, high-pressure technology enables pasteurization of foods with minimal effects on taste, texture, appearance, or nutritional value*, p. 38. Food Technology 11-08 • www.ift.org. Web.

Dr Sharma, Ranjan. "Market trends in high pressure processing (HPP) food." University of Qeensland, Seminar Presentation at the Centre for Nutrition & Food Sciences, Brisbane, 23 March, 2011. Print.

"Commercial Fish & Shellfish Technology; High Hydrostatic Pressure Processing Has Potential." Originally appearing in *Global Aquaculture Advocate*, Vol. 6:1 (February 2003). Journal Reprint.

Chapter 2

Pulsed Electric Field Processing

Background

Traditionally, pasteurization or heat processing is used to reduce the level of bacteria, spores, and other agents that cause spoilage of fruit juices, beer, milk, and other liquid foods. Heat processing, however, also reduces the flavour of these foods. Orange juice, for example, is susceptible to undesirable reactions during heat processing, resulting in off-flavours and dark colour. To avoid degradation of food quality during heat processing, researchers at Ohio State University (OSU), Washington State University, and many other institutions have conducted considerable research into non-thermal processing of foods. A highly effective non-thermal process is Pulsed Electric Field (PEF) Processing. In PEF processing, an electric field, rather than heat, chemicals, or irradiation, is used to kill microorganisms, spores, etc. Kill rates comparable to pasteurization have been demonstrated by multiple researchers across a wide range of liquid foods, including juices, milk, liquid eggs, and sauces.

The key to PEF processing is that the food stays cool because heat is not required for bacterial kill. Because there is little heating, the

taste and nutritional value of PEF treated foods are essentially indistinguishable from fresh (untreated) products. In PEF processing, a liquid food, such as a juice, sauce, or other pumpable product, is passed through a small treatment chamber, where it is subjected to a short (10 ns–20 μs) pulse of very high voltage. The high voltage field created across the liquid (approximately 35-50 kV/cm) kills microorganisms and spores by disrupting cell membranes. The pulses are so short and frequent that all of the liquid in a pipe can be treated as it flows through the treatment chamber. By using multiple treatment chambers to apply pulses to a stream of fluid, kill ratios of 5-9 log have been achieved, similar to pasteurization.

Multiple experiments have demonstrated that the shelf life of PEF processed food is comparable to that yielded by pasteurization. No adverse impact on the taste or nutritional value of the food was found. The primary ***limitation*** to commercialization of PEF processing has been the inability to generate the high voltage pulses having sufficient peak power (typically megawatts) needed to process large quantities of fluid economically.

Typical laboratory systems have used vacuum tubes or spark gaps to create the high voltage pulses. These systems are expensive, inefficient, and limited in either peak or average power. Without the ability to provide high peak power levels efficiently, the throughput and cost of these systems is not economical for commercial operations. The emergence of solid-state pulsed power systems, which can be arbitrarily sized by combining switch modules in series and parallel, removes this limitation. The solid-state pulsed power system delivered to OSU provides bipolar 60 kV, 600 A pulses at 75 kW average power into four PEF treatment chambers. It consists of two fully independent switching power supplies and solid-state switches: one at positive voltage, and the other at negative voltage. This provides a very high level of pulse flexibility to the OSU researchers as they optimize PEF treatment parameters across a wide range of liquid foods.

Introduction and Overview

The ever-increasing trend toward nutritionally qualified foods has challenged food technology to produce fresh-like foods by replacing thermal treatments with alternative methods of preservation. Thermal processing is a major technology that has been commonly used in the food industry to increase shelf life and maintain food safety with low processing costs.

To qualify as an alternative method, a new technology should have significant impact on quality while at the same time maintain the cost of technology within feasibility limits. In recent years, several technologies have been investigated that have the capability of inactivating microorganisms at lower temperatures than typically used in conventional heat treatments (Lado and Yousef, 2002). Therefore, non-thermal methods correspond to the expectations for minimally processed foods of fresh quality, which have higher nutritional value because of colour and flavour retention.

Among all emerging non-thermal technologies, high intensity pulsed electric fields (PEF) is one of the most appealing technologies due to its short treatment times and reduced heating effects with respect to other technologies. High intensity pulse electric field is highly appreciated as a non-thermal food preservation technology that involves the discharge of high voltage electric short pulses through the food product.

With the use of electric fields, PEF technology enables inactivation of vegetative cells of bacteria and yeasts in various foods. As bacterial spores are resistant to pulsed electric fields, applications of this technology mainly focus on food-borne pathogens and spoilage microorganisms, especially for acidic food products. In addition to the volumetric effect of PEF technology in controlling the microbiological safety of foods in a fast and homogenous manner, successful application provides extended shelf life without the use of heat to preserve the sensory and nutritional value of foods.

PEF technology has the potential to economically and efficiently improve energy usage, besides the advantage of providing microbiologically safe and minimally processed foods. Successful application of PEF technology suggests an alternative substitute for conventional thermal processing of liquid food products such as fruit juices, milk, and liquid egg (Bendicho et al., 2002a).

Pulsed electric field (PEF) processing is a method for processing cells by means of brief pulses of a strong electric field. PEF holds potential as a type of low temperature alternative pasteurization process for sterilizing food products. In PEF processing, a substance is placed between two electrodes, then the pulsed electric field is applied. The electric field enlarges the pores of the cell membranes which kills the cells and releases their contents. PEF for food processing is a developing technology still being researched.

There have been limited industrial applications of PEF processing for the pasteurization of fruit juices. Nearly all cells have pores which control the flow of nutrients and wastes into and out of the cell. Once the pores are sufficiently enlarged, the contents of the cell are released into the surrounding medium and foreign matter may enter the cell membrane. PEF produces products with slightly different properties than conventional pasteurization treatments. Most enzymes are not affected by PEF. The fact that the maximum temperature reached is lower than in thermal pasteurization means that some of the flavours associated with the raw material are not destroyed.

The lack of heat treatment makes PEF somewhat comparable to irradiation as a treatment. Spores, with their tough protective coats and dehydrated cells are often able to survive PEF processing. The survival of spores and enzymes means that products should be refrigerated after passing through PEF processing in order to slow the action of the enzymes and keep pathogens from growing.

The PEF process holds promise as a more efficient way of getting useful products out of cell membranes. PEF is particularly well suited to processing fruit and vegetable juices because the enlargement of the cell pores makes juice extraction easier. PEF may be useful in extracting sugar from sugar beets and oils from oil bearing plants. PEF may have a use in the developing field of extraction of oil and other products from microorganisms such as algae. Part of the potential utility of PEF is that the properties of the electric field can be modified to have differing effects on cells. Most plant cells can survive weak electric fields, but over around 15,000 volts per centimetre most plant cells are killed. At over 30,000 volts per centimetre, most other cells such as bacteria and fungi are also killed. Issues with PEF include the potential for electrical arcs and a temperature increase on the order of 30 degrees Celsius (54 °F). In liquid PEF processing gas bubbles tend to allow for electric arcing between the electrodes which will burn the substance being processed and potentially generate unwanted materials, such as carcinogens, in the processed material. One solution to gas bubble formation is to pressurize the liquid.

In static PEF for non-liquids there is a greater limitation on the maximum voltage possible without electric arcing through the air. This necessitates transferring the solid to an oil filled chamber before a high voltage PEF is applied. It has been shown that exposing biological cells to an electric field provides a potential to influence permeability of cell membranes and to induce structural changes and a local membrane breakdown. Based on this phenomenon called electroporation a broad variety of applications in food processing and bio-processing have been studied since the 1960s. In plant or microbial genetics an electroporation is used to induce foreign material such as DNA into cells. The process of reversible pore formation has to be perfectly controlled to maintain viability of the organisms during the application of the PEE This principle can also be utilized to induce stress reactions and secondary metabolites biosynthesis, which can be desirable food constituents.

Increasing treatment intensity an irreversible membrane breakdown of plant, animal, or microbial cell membranes can be achieved. Disintegration of biological cells and break of permeability barrier often is a crucial pre-treatment step to improve mass transfer rates. A PEF treatment can be utilized to substitute conventional disintegration techniques such as grinding, thermal, or enzymatic treatments prior to processes such as drying, extraction, or pressing. Pore induction by PEF can, dependent on treatment intensity, also cause a loss of vitality and be utilized to achieve an inactivation of microorganisms in liquid media. Several attempts to achieve an industrial exploitation have been made (Toepfl et al., 2005a), in 2006 the first commercial application of the technique has been reported (Clark, 2006).

Industrial implementation of pulsed electric field electro-technology (PEF) for food preservation has been rather slow, despite its potential to produce safe, nutritious, and high quality products. Several research groups around the world are in a race to validate and optimize the operation of PEF systems. Insufficient kinetic studies and inaccurate treatment delivery assessment are some of the main obstacles to the implementation of this technology. Equivalency among PEF systems, treatment delivery, and control variables need to be clearly defined before actual industrial implementation takes place. Limited commercial availability of PEF systems, mainly due to the complexity and high cost of pulses, and to technical and economical limitations of the scaling-up process, reflects into high initial investment and operation costs.

PEF imposes a strong electric field on a flowing fluid for a very short time. Above critical field strength of about 15,000 V/cm, vegetative cells are killed. In practice, higher field strengths are used, up to about 35,000 V/cm for disinfection (destruction of bacteria, fungi, and other microbes). Lower-strength fields are used for static PEF.

Present Status of Pulsed Electric Field Technology

The extent of improvement in a food processing company achieved by an emerging technology generally reflects the interest in that technology by the food industry. The method of high intensity pulse electric fields used to inactivate microorganisms has been under research for nearly 45 years. During this time span the technology has proved to be most effective in the inactivation of vegetative bacteria, yeasts, and moulds, while bacterial spores are much more tolerant. Many successful steps have been taken in the design of system components and inactivation mechanism for different species.

Inactivation kinetics and the effect of PEF on spores are some of the most discussed issues in recent studies. Methods applied to thermal processing technologies have been used to explain inactivation kinetics neglecting the deviations from linearity for these plots. The lethal effects of PEF processing on spores are still being explored. Additionally, the synergistic effect of PEF technology in combination with other mild preservation methods is one of the research interests popular in recent years.

Considering the effectiveness of PEF treatment on liquid products, such as milk, fruit juices, liquid egg, and any other pumpable food products, extensive research has been done to implement the process at an industrial level. Flavour freshness, economic feasibility, improvements in functional and textural attributes and extended shelf life are some of the main points of interest besides achievement of microbiological safety of food products (Dunn, 2001). Among all liquid products, PEF technology has been most widely applied to apple juice, orange juice, milk, liquid egg, and brine solutions.

The rapid increase in the market segment of fruit juices has led manufacturers to seek ways to overcome the thermal effects of processing while increasing the shelf life of juices (Sadler et al., 1992; Hodgins et al., 2002). Application of PEF is especially promising for the citrus industry. PEF processing has been successful in a variety of fruit juices with low viscosity and electrical conductivity such as orange, apple, and cranberry juice. Additionally, the colour change in fruit juices was reportedly less in juices treated by PEF (Yeom et al., 2000).

Studies conducted on the effects of PEF on dairy products such as skim milk, whole milk, and yogurt compromise a major section of PEF applications (Alvarez and Ji, 2003). Chronological developments in pulsed electric field processing were appended in Table 2-1. In addition to inactivation of microorganisms and enzymatic treatments, the application of PEF provides highly effective pre-treatment of foods and food ingredients in an ecological and gentle way (Guderjan et al., 2005). The ability of pulsed electric fields to destroy cell membranes has been used in several applications (Lebovka et al., 2003). The effect of combined pressure-PEF treatment enhances the solid-liquid expression of different biological tissues and increases the juice yield (Wang and Sastry, 2002; Vorobiev et al., 2004).

Table 2-1 ***Chronological Developments in Pulsed Electric Field Processing***
(Source: Raso and Heinz, 2006)

Years	Development
1935	*Application of electric fields to preserve foods first appeared with the Electro-pure method for pasteurization of milk.*
1940	*Electric fields were being used in the food processing process for purposes other than inactivation of microorganisms.*
1950	*The electro-hydrolytic method was introduced to inactivate microorganisms suspended in liquid systems.*
1960	*A variety of PEF equipments/methods was described, including non-thermal effects of pulsed electric fields on microbes. A systematic study to assess the non-thermal bactericidal effects induced by electric fields. Along with, a series of papers were published on the use of electrical field technology.*
1970	*Researchers in the area of genetic engineering and medical science began exploring the action mechanism at a cellular level. Introduction of the dielectric rupture to promote cell fusion, the action mechanism of electrical pulses, combining the theory of membrane disruption by electric fields with the previous work on process parameters, sensitivity of microorganisms, and inactivation mechanisms.*
1980	*Commercialization of PEF applications. Different from all existing technologies, exploration of PEF technology, unknowns of pulsed electric fields, patents on PEF, on the microbial inactivation, tissue response to electric fields, enzyme inactivation, engineering aspects, inactivation kinetics, modelling, and scale-up. Industrial applications of pumpable electrically conductive media.*
1990	*Promising treatment for fluid foods such as orange juice and milk, microbial reduction, disruption of vegetable and animal cell membranes subjected to pulsed electric fields, pulsed electric field processing systems for antimicrobial treatment of liquids and pumpable foods.*
2000	*Achievement in the commercialization of PEF applications. Currently, commercial PEF systems are available that include both bench-top and industrial systems, besides many different lab-scale and pilot-scale PEF systems. The use of solid-state pulsed power systems, with reliability and process consistency in PEF processing when used in a laboratory scaled up to commercial food processing applications.*

PEF technology has recently been used in alternative applications including drying enhancement, enzyme activity modification, preservation of solid and semisolid food products, and waste water treatment, besides pre-treatment applications for improvement of metabolite extraction. The ability of PEF can be successfully used to enhance mass and heat transfer to assist drying of plant tissues. Studies conducted on different plant tissues such as potato tissue, coconut (Ade-Omowaye et al., 2000), carrots (Rastogi et al., 1999), mango (Tedjo et al., 2002), and apple slices (Ade-Omowaye et al., 2002) reported increased yield of water removal by 20-30% when exposed to low intensity electric fields.

Pre-treating red peppers at the PEF-enhanced initial drying rate significantly indicated that combining PEF and partial osmotic dehydration in solutions before air drying may offer good potential in satisfactorily enhancing mass transfer rates and preserving the colour quality of red peppers (Ade-Omowaye et al., 2003).

As stated before, PEF technology commonly focuses on processing pumpable and homogenous liquid foods free of particles and air. In the case of solid foods the mixture is mixed with air, contributing to the low electrical conductivity of the product, hence, not limiting the maximum applicable electric field intensity. Some studies conducted on model foods, viscous foods such as yogurt and rice pudding (Ratanatriwong et al., 2001), or particulate foods such as pea soup with plastic beads (Dutreux et al., 2000) reported successful results for PEF applications on solid or semisolid foods. Patents on pulsed electric field technology were detailed in Table 2-2

Table 2-2 ***Patents on Pulsed Electric Field Technology***

(Source: Sepulveda and Barbosa-Canovas, 2005)

References	Patents
Held and Chauhan(2002)	*Destruction of waste-activated sludge*
Morshuis et al. (2002)	*Device and Method for Pumpable Food Products*
Heesch (2002)	*Pulse Electric Field Treatment System*
Zhang and Qiu (2001)	*High Voltage Pulse Generator*

Lelieveld and Volanschi (2001)	*Method and Apparatus for Preserving Food*
Mastwijka nd Bartels (2001)	*Integrated Modular Design for Treatment Chamber*
Bushnell (2000)	*A Serial Electrode Treatment Cell for Pumpable Food*
Mittal et al. (2000)	*Method and Apparatus for Electrically Treating Food Stuff*
Addeo (2000)	*Use of PEF Coupled With Rotational Retorting*
Bushnell et al. (2000)	*Uniform Product Flow in a PEF Treatment Chamber*
Qin et al. (1998)	*Continuous Flow Electrical Treatment Chamber*
Hayden (1998)	*Method of Inactivation in Liquids*
Yinet al. (1997)	*High Voltage PEF Chambers for Liquid Products*
Qin et al. (1997)	*Continuous Flow Electrical Treatment Chamber*
Zhang et al. (1996)	*Batch Mode Food Treatment Using PEF*
Bushnell et al. (1996)	*Process for Inactivation of Microorganisms With PEF*
Bushnell et al. (1995a, b)	*Prevention of Electrode Fouling, and Electrochemical Effects in PEF Systems*
Bushnell et al. (1993)	*PEF Systems for Extension of Shelf Life*
Bushnell et al. (1991)	*PEF Systems for Extension of Shelf Life*
Dunn et al. (1991)	*Methods for Preservation by PEF*
Doevenspeck (1991)	*Electric Impulse Method and Device for Treatment*
Dunn et al. (1989)	*Methods for Preservation with PEF*
Dunn and Pearlman (1989)	*Device for Shelf life Extension for Liquid Foods*
Dunn and Pearlman (1987)	*Methods for Shelf Life Extension of Liquid Foods*

Principles and System Components

The high intensity pulsed electric field processing system is a simple electrical system consisting of a high voltage source, capacitor bank, switch, and treatment chamber (Figure 2.1). Generation of pulsed electric fields requires a fast discharge of electrical energy within a short period of time. This is accomplished by the pulse-forming network (PFN), an electrical circuit consisting of one or more power supplies with the ability to charge voltages (up to 60 kV), switches (ignitron, thyratron, tetrode, spark gap, semiconductors), capacitors (0.1-10 μF), inductors (30 μH), resistors (2 Ω-10 MΩ), and treatment chambers (Gongora-Nieto et al., 2002). Typical values of industrial power modulators for the treatment of liquid foods and other biological liquids or aqueous materials were given in Table 2-3 below.

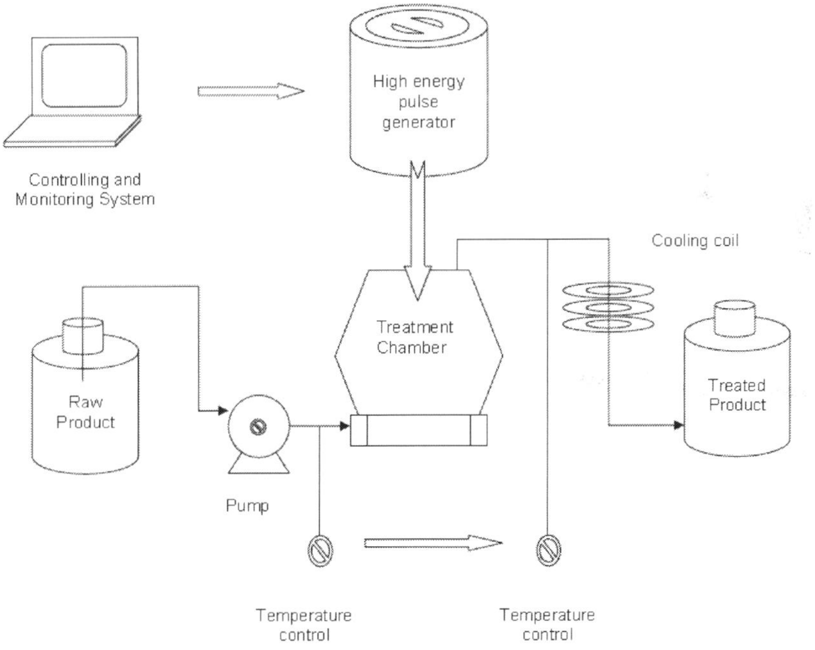

Figure 2.1 ***Flow Chart of a PEF Food Processing System With Basic Components***
(Source: European Journal of Food Research & Review, 1(2) (2011) : 71-93)

Table 2-3. ***Industrial Power Modulators for the Treatment of Liquid Foods and Other Biological Liquids or Aqueous Materials Should Have the Following Typical Values***

(Source: Raso and Heinz, 2006)

Specific Electrical Data for Medium Treatment	Specific Data for Facilities
Specific conductivity of the medium	***Mass flow***
• range: 10–2–10 S/m	• range: 5–300 t/h
• typical value: 0.3 S/m	• typical value: 20 t/h
Electric field strength in the medium	***Average power***
• range: 1–100 kV/cm	• range: >20 kW
• typical value: 30 kV/cm	• typical value: 300 kW
Power density	***Treatment time***
• range: 2 MW/L-30 GW/L	• range of single pulse durations: 1–1000 μs
• typical value: 3 GW/L	• range of pulse rates: 1–3000
Specific energy consumed in the medium	• typical value of treatment time (rectangular pulses): 10 μs
• range: 0.5–5000 MJ/t	***Pulse repetition rate***
• typical value: 50 MJ/t	• range: 1–2000 pps
	• typical value: 1 pps (remark: this value yields for laboratory devices; for industrial applications higher repetition rates in the order of ~ 100 pps are required).

Working of Pulses Electrical Field

When exposed to high electrical field pulses, cell membranes develop pores either by enlargement of existing pores or by creation of new ones. These pores may be permanent or temporary, depending on the condition of treatment. The pores increase membrane permeability, allowing loss of cell contents or intrusion of surrounding media, either of which can cause cell death. PEF has limited effects on spores and only appears to affect a few enzymes. Enzymes are important in juice processing because surviving enzymes can reduce pectin, which then can be less effective in keeping fruit particles suspended. Some sedimentation is common in fruit juices, but too much is unattractive.

Some surviving enzymes may also enable discoloration and production of off-flavours. PEF offers a 5-log reduction of most pathogens and is considered a pasteurization process, so products must be refrigerated. An important process consideration is prevention of post-process contamination, so filling is in an aseptic or clean room environment and containers must be cleaned and possibly sterilized. The perforation of cell membranes caused by PEF also applies to fruit and vegetable cell walls, so a potentially beneficial side-effect of the process is improved extraction of juice from cells. This phenomenon is also applied in another promising use, concentration of sewage sludge, a suspension of live and dead cells and organic matter which can be very hard to filter and concentrate. PEF, by killing live cells and reducing their ability to retain water, greatly improves filtration. Extraction of sugar from beets and starches from potatoes may also be improved by PEF.

Process Variables

Important process variables include the electric field, which can have various wave forms, strengths, and distribution in the treatment chamber; temperature, pressure, time of exposure, and high power sources.

Electric Field

The electric field is generated by equipment similar to that used in radar. This has some consequences for cost and availability. The most typical equipment generates a short square wave and reverses polarity, in part to avoid erosion of electrodes. However, a bipolar generator costs about twice as much as a mono-polar one. Other wave forms include exponential decay and sinusoidal. The sinusoidal form is somewhat easier to generate, as it uses equipment similar to common radio equipment, but it reaches its peak power only for an instant and so delivers less energy per cycle above the critical field strength than does a square wave (Figure 2.2).

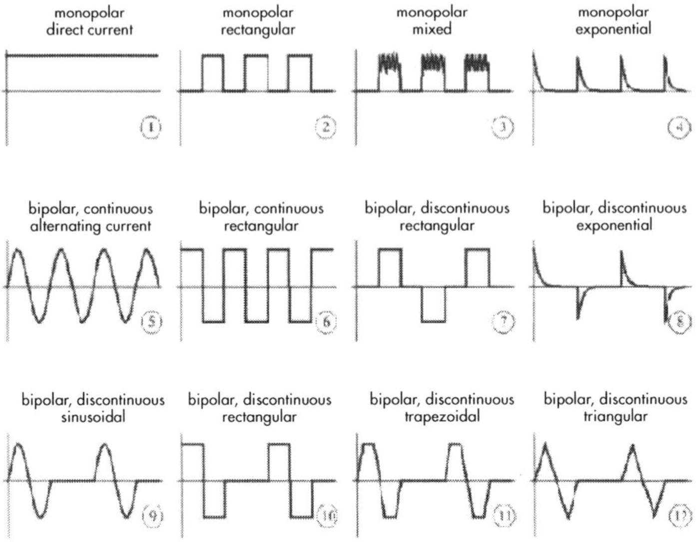

Figure 2.2 ***Typical Voltage Waveforms for PEF Technology***
(Source: European Journal of Food Research & Review, 1(2) (2011) : 71-93)

Temperature

There is some temperature rise across the PEF treatment chambers due to the delivery of electrical energy, but the preservative effect is primarily non-thermal. A typical temperature change is about 30 °C for orange juice and less for apple juice Processes typically operate at 35–50 °C because it has been found that microorganisms are more tolerant at low temperatures.

Pressure

Pressure is applied to inhibit the formation of air bubbles in which electrical arcing could occur with fields above 20,000 V/cm.

Time of Exposure

The field is cycled about 1,000 times/sec, and the fluid is exposed to multiple pulses by passing it through several treatment chambers. The treatment chambers are separated from each other by insulators, which may be ceramic or polymer. Each chamber has two electrodes, which are made from a conductor or semi-conductor. Options for chamber design include parallel plates, co-axial, and co-field. PEF may also be applied in a static (non-flowing) approach for foods such as meats.

High-power Sources

Several methods exist for generating pulsed electric fields. Depending on the specified type of voltage wave form, as well as on the available elements (capacitors, inductors, transformers, switches), the possibilities vary between simple circuits and very sophisticated networks. Basically, these circuits can be distinguished as being:

- *Basic pulsed power circuits (Fig. 2.3)*
- *Circuits with voltage multipliers*
- *Pulse forming circuits or*
- *Networks with pulse forming switches.*

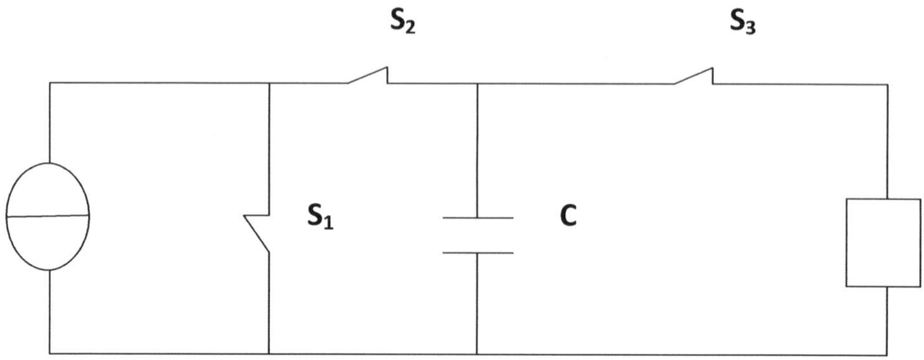

Figure 2.3 ***Example of a Capacitative Circuit***
(Source: European Journal of Food Research & Review, 1(2) (2011) : 71-93)

Applications of PEF Technology For the Food Industry

The application of PEF in food processing gained considerable attention within the last decades, utilizing its impact on cell membranes. Apart from preservation the disintegration of biological tissue is often a key step in food processing prior to winning of intracellular compounds. It is noteworthy that an electro-permeabilization can be performed continuously and in a time-scale of seconds, the treatment, therefore, can easily be implemented into existing processing lines.

Juice Processing

Applying PEF to cellular tissue an increase in mass transfer coefficients was observed due to cell membrane permeabilization. Based on this effect a PEF application can replace or substitute conventional techniques in fruit and vegetable juice processing. First, attempts to apply PEF technology for disintegration of cells have been based on 10-12% increase of juice yield when applying electroplasmolysis to apple tissue. Juice expression of apple under simultaneous mechanical pressing and PEF treatment, a significant increase in juice yield was reported after a treatment at 520 V/cm and 100 μs treatment time.

For carrot juice an increase of juice yield from 60.1 to 66.4% was found in comparison to an untreated sample; in the same way the dry matter of the sample was increased from 13 to 15% (Knorr et al., 2001). The influence of a PEF pre-treatment of apple mash was investigated and a juice yield of 83.0% was obtained. A PEF treatment of grapes resulted in a juice yield of 87%. The beneficial impact of a PEF treatment for wine production was also reported by Sigler et al.

It was shown that the press ability of juices from soft vegetables such as potato, after a PEF treatment was improved. Application of PEF for juice processing provides a potential to replace or enhance conventional cell disintegration techniques In the context of consumer demand for functional food with a composition close to fresh products and a high content of physiologically valuable compounds an increased extractability of intracellular compounds such as pigments or phenolic substances could provide an enormous potential for product development.

A system with a peak voltage of 20 kV, an average power of 7 kW and a production capacity of 2 ton/h has been developed for the treatment of fruit mashes. This provides a potential to extract high quality pectin from the pomace after juice winning and therefore a step toward a more economic and sustainable processing.

Meat and Fish Treatment

Disintegration of animal cellular tissue might also be utilized to enhance processes where an uptake of substances is required, such as marination or curing of fish or meat products. Salt or a combination of salt and antimicrobial agents such as nitrite is often applied for preservation; spices are added to develop product-specific taste and aroma profiles. In case of raw ham a long-term curing and air drying is applied. During such procedures a PEF treatment can be applied to improve mass transfer processes and to accelerate curing, reducing the time requirements.

Not many research works is available in this field of PEF application, but recently the interest in the potential of PEF application for meat and fish products increased. In contest to plant material or liquid food the effect of a PEF treatment on protein-based, solid food structures such as meat and fish tissue or protein gels has been studied. (Sitzmann and Munch, 1988; Hafsteinsson et al., 2000; Barsotti et al., 2001; Gudmundsson and Hafsteinsson, 2001).

Research work on feasibility of a PEF treatment in meat processing shown that the fermentation of raw sausages (Salami type) can be accelerated by improving the availability of intracellular liquid for fermenting cultures. During meat processing salt as well as sodium nitrate are applied for preservation, reduction of water activity as well as meat curing. The impact of a PEF treatment on pork muscle is was investigated by electron micrographs, showing the formation of a swollen, sponge-like tissue structure (Figure 2.4).

Removal of water to preserve food products such as fruits or vegetable accounts for a significant amount of energy costs in food processing. The electro-permeabilization of cell membranes, leading to a drastic increase in mass transfer rates, can be utilized to enhance drying rates of cellular tissue. The drying rates of osmotic drying and the diffusion coefficients of carrots were investigated by Rastogi et al.

After a PEF treatment at field strength of 1-2 kV/cm the time required for fluidized bed drying of potato cubes was reduced. For apple slices an increased osmotic drying rate and improved rehydration capacity as well as shorter rehydration times were reported (Taiwo et al., 2002).

Ade-Omowaye et al. (2001) investigated the relationship between a PEF treatment and drying rates of plant-based foods during osmotic dehydration and air drying. As mass transport across the cell membrane is influenced by a PEF treatment, it can be utilized to significantly increase mass transfer rates by 10-30% when applying 5-20 pulses at field strength of 1.0 kV/cm. An increase in mass transfer rates, resulting

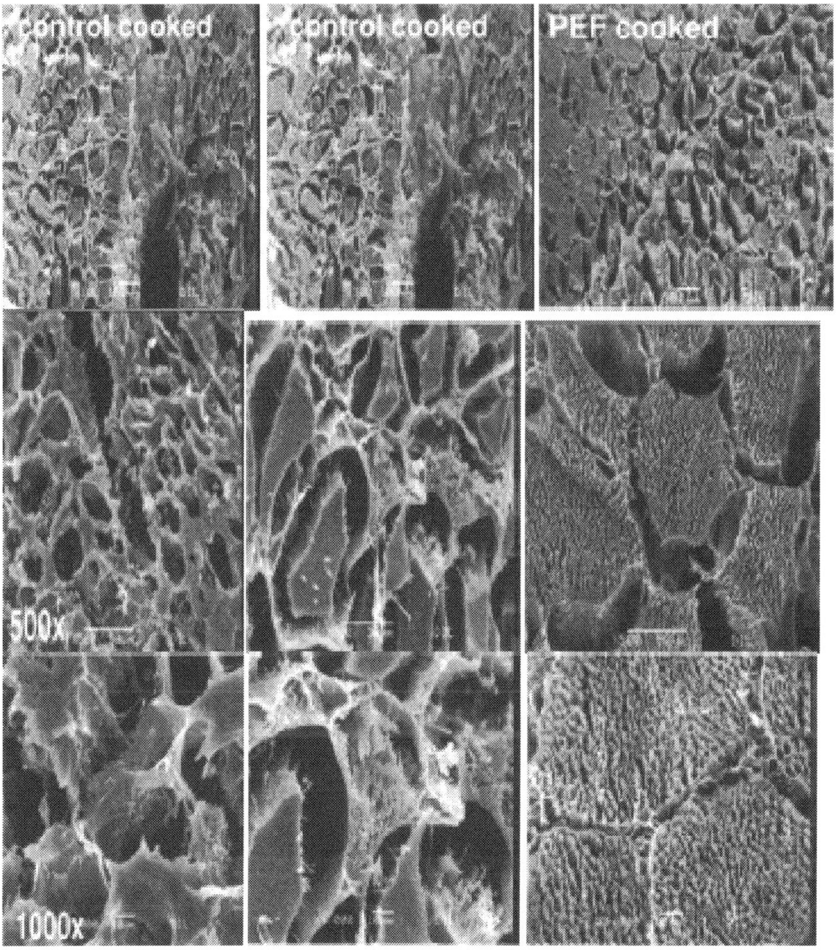

Figure 2.4 ***Electron Micrographs of Pork Meat Untreated (left column), After Cooking (centre column) and After PEF-treatment and Cooking (right column)***
To all samples an injection of 20% brine containing 10% of pickling salt was applied, cooking was performed at 75 °C, 90% humidity until a core temperature of 64 °C. (Source: European Journal of Food Research & Review, 1(2): 71-93, 2011)

in faster water transport to the product surface and therefore reduction of drying time after a pre-treatment will lead to drastic saving of energy and better utilization of production capacities during convective air drying. Taking into account the low energy input required for a PEF treatment of plant or animal tissue (2-20 kJ/kg), it is evident that there is a potential to reduce the total energy input for product drying.

Sugar Processing

Conventional procedures for production of sugar from beets involve an extraction at elevated temperatures (68-72 °C) after carving the fruits into cassettes. For disintegration and destruction of cell membranes a thermal treatment at temperatures in the range from 70 to 78 °C is applied. The membrane denaturation results in an acceleration of sugar release into the extraction media, but also cell wall components such as pectin may become soluble and can diminish juice purity and quality. In addition, the thermal denaturation, as well as the hot water extraction require a significant amount of energy, as high as 175 kJ/kg of treated beet (Schultheiss et al., 2002).

It has been shown that mechanically pressed, raw juice has a higher sugar concentration and contains less nonsugars, but juice yield remains unacceptable (Bliesener et al., 1991). A PEF treatment of sugar beets prior to extraction could increase mass transfer rates and could allow to reduce extraction temperatures or to apply mechanical pressing. The applicability of a PEF pre-treatment prior to an extraction at ambient temperature has been investigated by Eshtiaghi and Knorr (2002). It was shown that after a PEF treatment at 2.4 kV/cm and a pulse number of 60, similar cell disintegration than after a thermal treatment at 75 °C for 15 min was obtained. A three-step pressing at a pressure of 5 MPa and intermediate addition of water was suggested to achieve a high sucrose content juice after a short processing time of 30 min in comparison to up to 90 min for thermal extraction. The energy input required was 12 kJ/kg.

Bouzrara and Vorobiev (2000) reported a juice yield of 78% after application of 1000 pulses with a peak voltage of 1.2 kv. The specific energy input required was in the range of 100 kJ/kg. From an engineering point of view this high energy input could be significantly reduced by selection of appropriate processing parameters.

Schultheiss et al. (2004) reported an increase of juice yield by a factor of 2.1 after a PEF treatment in comparison to an untreated sample. Raw juice quality was maintained or improved even when using low quality beet as raw material. A comparison of juice quality to juice from thermally treated beet revealed superior juice quality. These results indicate the tremendous potential of PEF application for sugar processing. Prior to large industrial scale application of PEF as required for sugar processing the development of systems with production scale in a range of 10-20 ton/h and their successful exploitation will be required to confirm the technical feasibility.

Effectiveness of PEF Treatment in Food Preservation

The effectiveness of PEF treatment depends on many previously described factors. Authors report the effectiveness of PEF for specified processing conditions. Below there are described some examples for that aspect of food preservation. The inactivation of *Escherichia coli* by PEF was studied in liquid, solid, and semisolid foods or model systems by Manas et al. (2001). They found that agitation of the inoculated liquid samples (16 mmol/L sodium phosphate buffer) during pulse processing resulted in efficient microbial inactivation–5 log cycles at 33 kV/cm and about 25 °C after 261 μs of cumulated pulses. The highest extent of *Listeria innocua* inactivation in liquid whole egg (LWE) was 3.5 log cycles for an electric field intensity of 50 kV/cm, 32 pulses and total time duration of 64 μs (Calderon-Miranda, 1999).

Table 2-4 ***Parameters of PEF and Effectiveness of Microbial Inactivation in Different Liquid Food Products***

(Source: Vega-Mercado, 1997)

Product	PEF Parameters				Effectiveness of Microbial Inactivation	
	Electric Field Strength E (kV/cm) strength E (kV/cm)	Electric Field Strength E (kV/cm)	Number of Pulses n (-)	Tempperature of Product After PEF Treatment T (0C)	Microflora	Log cycles D (or description)
Orange Juice	6.7	20	5	45–50	*Saccharomyces cerevisae*	5D
Milk	28.6	100	23	22.8	*Escherechia coli*	3D
Milk	36.7	100	40	63	*Salmonella dubin2D*	3D
Milk	22	20	20	45–50	*Lactobacillus bulgaricus*	4.6D
Yoghurt	23-38	100	20	63	*Streptococcus thermiphilus*	2D
Liquid Whole Egg	25.8	4	100	37	*Escherechia coli*	6D
Skim Milk	45	1.8-6	64	35	*Escherechia coli*	2D
Fluid Food	12–25	1–100	25	45-55	Natural	Shelf life extended from 3 to 7 days

The presence of 37 IU nisin/ mL in LWE under the same PEF conditions resulted in a better effectiveness of *Listeria innocua* inactivation (4.4 log cycles). It points to better results achieved upon the application of combined methods for food preservation, which was confirmed in the next studies where liquid whole egg (LWE) with 0.15% addition of citric acid and after PEF treatment (E = 30 kV/cm, t = 489 ms, W = 6331 J/mL) had the shelf-life of 20 days at 4 °C. The same LWE but with 0.50% addition of citric acid and after a slightly different PEF treatment (E = 30 kV/cm, t = 55 ms, W = 357 J/mL) was characterized by 30-day shelf-life at a temperature of 4 °C (Gongora-Nieto et al., 2003).

The above examples indicate that a skilful combination of preservation methods results in a safer product with extended shelf-life. Although some studies have concluded that PEF preserves the nutritional components of the food, effects of PEF on the chemical and nutritional aspects of specific foods should be better recognized. The effectiveness of PEF treatment on different microorganisms and food systems was shown in Table 2-4.

The maximum microbial reduction of the presented examples was 6 log cycles for *Escherichia coli* in liquid whole egg. We have also evaluated the efficiency of pulsed electric field (PEF) against *Escherichia coli* contaminating the liquid whole egg (LWE) in our studies (Malicki et al., 2004). The samples of LWE were inoculated with the test bacteria and subsequently treated for 30 μs by the different number of pulses (20–180) of PEF (32.89 kV x cm-1).

The application of PEF resulted in a statistically significant reduction of the test microorganisms, proportional to the number of pulses used. Depending on the studied strain, the treatment with 150-160 PEF pulses was required to obtain the reduction of initial bacteria level by 4 log units. Considering the obtained results, PEF seems to be an effective technique that improves the microbiological status of LWE. Its industrial application is, therefore, highly advisable.

Moreover, we have studied functional and rheological properties of LWE after the PEF treatment (Oziembłowski et al., 2005). LWE was obtained from eggs of 27-week-old layer hens, Tetra SL. Parameters of the treatment were chosen according to earlier experimental results where a significant reduction of microflora was observed at

32.89 kV/cm using 20, 60, 100 pulses. It was concluded that the functional properties of liquid whole egg after PEF were not worse than those of the control sample.

Furthermore, foam ability and emulsifying capacity of LWE after PEF were significantly better with an increasing number of impulses compared to the control sample. Viscosity of LWE at a shear rate 250 (1/s) was higher after the PEF treatment: 112 mPa s (20 pulses) and 106 mPa s (100 pulses). For the control sample, the apparent viscosity was 102 mPa s. The results obtained indicate that pulsed electric fields could be useful in the preservation of the liquid whole egg, especially in the context of functional and rheological properties. It is obvious, however, that the sufficient improvement of microbiological safety was the main assessment criterion of that method.

References

Oziembłowski, Maciej, and Kopeć, Wiesław. "Pulsed electric fields (PEF) as an unconventional method of food preservation." *Polish Journal of Food and Nutrition Sciences*, Vol. 14/55, SI 1,(2005) : 31-35. Print.

Gaudreau, M. P. J.; Hawkey, T.; Petry, J.; and Kempkes, M. *Pulsed power systems for food and wastewater processing.* Diversified Technologies, Inc., USA. Print.

Singh1, Ranjeet and Kumar, Ashok. "Pulsed Electric Fields, Processing and Application in Food Industry." E*uropean Journal of Food Research & Review* 1(2) (2011) : 71-93. Print.

Siquan Li, M.S. "Pulsed electric field processing of functional foods." PhD Dissertation, The Ohio State University, (2003). Print.

Barbosa-Cánovas, Gustavo V. "Transferencia exitosa del conocimiento: Nuevas tecnologías para procesar alimentos." Centre for Non thermal Processing of Foods Washington State University, (2008). Print.

Chapter 3

Food Irradiation

Food irradiation is one of the most extensively and thoroughly studied methods of food preservation.

—Thakur and Singh (1994)

Historical Perspective

Early in the 1920's, a French scientist discovered that irradiation could be used to preserve food. This technology was not adopted in the US until World War II. At this time there was a need to feed millions of men and women in uniform. The US Army sponsored a series of experiments with fruits, vegetables, dairy products, fish, and meats.

In 1963, the US saw its first approval of food irradiation when FDA approved its use to control insects in wheat and wheat flour. In 1964, additional approval was given to inhibit the development of sprouts in white potatoes. In 1983, approval was granted to kill insects and control microorganisms in a specific list of herbs, spices, and vegetable seasonings. (The approved list of food products has been increased with later changes.)

Then, in 1985, treatment of pork to control trichinosis was added to the list of approvals. In the same year, approval was granted to control insects and microorganisms in dry enzyme preparations used in fermentation-type processes. In 1986, approval was granted to control insects and inhibit growth and ripening in such foods as fruits, vegetables and grains. Approval was granted in May 1990, for irradiation of packaged fresh or frozen uncooked poultry. The American FDA supports it as an effective control of microorganisms responsible for a major portion of foodborne illness, including *Salmonella*, Yersinia, and Campylobacter. This approval was the first approved process to "pasteurize" solid foods, such as poultry. The process reduces but does not eliminate all bacteria. Therefore, processed poultry are safe longer than unprocessed poultry, but still require refrigeration.

The latest approval for irradiation in the US was granted in December 1997, for red meat. A maximum dose of 4.5 kg/y was approved for uncooked, chilled red meat and meat product, and 7.0 kg/y for frozen red meat and products. These approvals are for the purpose of controlling microorganisms, including pathogens such as *E. coli* 0157:H7. Just as with irradiated poultry, irradiated red meats still require refrigeration or freezing. Higher doses that sterilize frozen and packaged meats were approved in 1995 for use by NASA only. Legislation to allow the irradiation of seafood is being developed.

Principles and the Process

The energy employed in food irradiation technology is referred to as ionizing irradiation. Ionizing radiation occurs when one or more electrons are removed from an atom. Electrons orbiting at minimum energy level or ground state can be raised to higher levels, becoming electronically excited (excitation). If enough energy is transferred to an orbital electron, the excited electron may be ejected from the atom (ionization). A minimum amount of absorbed energy, called ionization potential, exists for each electron energy level necessary to exit the atom domain. If the energy absorbed by the electron is greater than its ionization potential, the excess energy enters a kinetic state, enabling the electron to leave the atom domain. Although each electron behaves individually, electrons can be used in large numbers, called electron beams, to irradiate food.

Electron beams are produced from commercial electron accelerators. One advantage of electron beam radiation is that the electron accelerators can be switched off when not in use, leaving no radiation hazard; however, the penetration of electron beams into foods is limited. Electrostatic forces tend to attract charged particles such as electrons (negatively charged), limiting the electron beam penetration into foods. Therefore, electron beam radiation is limited to small food items, such as grains, or to the removal of surface contamination of prepared meals.

Photons of electromagnetic radiation such as those created by x-rays and gamma rays travel without charge; thus, they do not interact with electrostatic forces while travelling through the food, and can penetrate deeper than electron beams. One mode of producing x-rays is through the use of electron beams. If the highly accelerated electrons penetrate a thin foil of certain metals, such as tungsten, tantalum, or any other material capable of withstanding high heat, x-rays are produced.

Radioactive isotopes such as cobalt 60 or cesium 137 are used to produce gamma rays. Although radiation isotopes cannot be switched off, they have the advantage of producing gamma rays, which offer the largest penetration among all ionization sources. Dose is the most important parameter in food irradiation. The quantity of energy absorbed by the food is measured in grays (Gy). One gray equals one joule per kilogram of matter. A gray (equal to 100 rad) is a very small quantity; therefore, the dose is expressed in kilograys (kGy). Up to 10 kGy is still a very small amount of energy, equal to the amount of heat required to raise the temperature of water 2.4 °C. For this reason irradiation is considered a "cold" or non-thermal technology

The processes can be explained as follows:

- *There are several processes that are collectively referred to as "food irradiation". Food is irradiated by placing it in, or moving it through, a field of ionizing energy consisting of electron beams or gamma rays or x-rays. Irradiation produces a wide range of beneficial effects on various foods including pathogen reduction, disinfestations of insects, growth inhibition, control of parasites, and shelf life extension.*

- *There are several processes that are collectively referred to as "food irradiation". The object of each process is to kill or impair the breeding capacity of unwanted living organisms or to affect the product morphology in a beneficial way that will extend shelf-life. Each process has an optimal dose of ionizing energy (radiation) dependent on the desired effect. The dose of radiation is measured in grays (Gy). A "gray" is a unit of energy equivalent to 1 joule per kilogram. This unit of measure is based on the metric system. Thus, 1 kilogray (kGy) is equal to 1,000 grays (Gy). All three forms of ionizing energy have the same effect, gray for gray.*

Some of the major processes are:

Pasteurization (Pathogen Reduction) Irradiation is used to effectively eliminate disease causing organisms including bacteria and parasites. (e.g., irradiating ground beef to make it safe from *E. coli* O157:H7. Irradiating live oysters to make them safe from *Vibrio*.)

Sterilization Irradiation is used at a very high dose to eliminate all organisms so that refrigeration is not required (shelf stable). (e.g. Certain foods are sterilized for NASA astronauts).

Sanitation Irradiation is widely used to reduce organisms for spices, herbs, and other dried vegetable substances. (e.g., irradiating spice blends that are added to meat for hot dogs and other "Ready to Eat" products that may not be cooked again).

Shelf-life Extension Shelf-life can be extended for certain foods using radiation by lowering the population of spoilage causing organisms, including bacteria and mould. On certain fruits and tubers, irradiation delays ripening and/or sprouting. (e.g., irradiating berries to reduce mould. Irradiating fresh fruits to extend their market reach.

Irradiating potatoes, onions, and garlic to impair cell division and hence allow them to go through the "off" season without sprouting).

Disinfestation Irradiation is used to stop reproduction of both storage and quarantine insect pests. (e.g., irradiating foreign produced mangoes to eliminate the seed weevil, which is a quarantined pest, for import to the US. Irradiating papaya to eliminate fruit flies, which are quarantined pests, for import from Hawaii or foreign countries into the US mainland).

All three forms of irradiation are referred to as a "cold process". Although all of the radiation energy is converted to heat during treatment, the process typically increases the product temperature by about 1 degree Celsius.

Equipment

Food is irradiated in "irradiators" that use electron beams or gamma rays or x-rays as their source of ionizing energy (radiation). Irradiators are designed to enable the irradiation of the food products to the desired dose and dose uniformity, without exposing workers or members of the public to radiation and without any effect on the environment. Food is irradiated in "irradiators" that use electron beams or gamma rays or x-rays as their source of ionizing energy (radiation).

All commercial irradiators have four primary components, a source of radiation, a method of product conveyance, "shields" to prevent exposure of personnel and the environment to radiation and safety systems. Ionizing radiation is penetrating energy and thus, products are usually irradiated after they are fully packaged. Below is a description of the four types of irradiators that are commercially available or in use today for food processing. The choice of which irradiator is most cost effective for a particular product depends on the type of product, how it is packaged, the product dose, dose uniformity requirements, and, most important, logistics.

Electron Beam Irradiator (employing a radiation chamber) The source of electron beams is an "accelerator." Accelerators generate and accelerate electrons very fast towards the food product being irradiated. Because electrons have mass, they can only penetrate about 1.5 inches (3.8 cm) into a typical food product or about 3.5 inches (8.9 cm) if the food product is irradiated on both sides. Electrons also have an electric charge. This charge allows the stream of accelerated electrons to be scanned by magnets to track across the product.

A commercial food electron beam irradiator accelerates the electrons to energy of up to 10,000,000 electron volts (10 MeV). Electron beam irradiators typically use massive concrete, steel or lead shielding. Electron Beam accelerators can be turned on and off. Safety interlocks ensure that a person cannot enter the radiation chamber where the food is being irradiated when the accelerator is "on." Product is usually passed through the scanned "beam" on roller type conveyors.

Gamma Irradiator (employing a radiation chamber) The source of photons in a gamma irradiator is cobalt-60. Unlike electron beams that are generated on site using electric power, cobalt-60 is produced off site in nuclear reactors and transported in special shipping containers ("casks") to the site.

Cobalt-60 is a solid radioactive metal that is contained in two welded encapsulations of stainless steel creating a "sealed source." The sealed source contains the "radioactive" Cobalt-60, but allows the photons ("radiation") to pass through the encapsulations and ultimately into the food product. Because Cobalt-60 photons have no mass, they can penetrate more than 24 inches (60 cm) of food product if irradiated on both sides.

Gamma irradiators that employ a radiation chamber typically have shields made out of massive concrete or steel. Cobalt-60 continuously emits radiation and cannot be turned "off". To allow personnel access to the chamber, the source is lowered into a storage pool of shielding water when it is not being used to irradiate product. The shielding water does not become radioactive. Safety interlocks are used to assure that a person cannot enter the chamber when the source is not in the stored position (at the bottom of the pool of water). Hanging carriers, totes and roller conveyors are typically employed to move the product through the chamber.

Gamma Irradiator (underwater) Like the radiation chamber irradiator above, an underwater gamma irradiator uses Cobalt-60. Unlike a radiation chamber irradiator, an underwater irradiator stores the Cobalt-60 permanently at the bottom of a pool of water. Instead of raising the Cobalt-60 into a shielded chamber, the product, placed in water free containers, is lowered to the bottom of the pool adjacent to the Cobalt-60 to receive a dose of radiation. The water acts as the shield. The shielding water does not become radioactive. No above ground shielding or radiation chamber is present. There is no need for interlocks to prevent personnel from entering a radiation chamber when the Cobalt-60 is present, because there is no radiation chamber. Typically, the product is loaded into water free containers and the containers are lowered/raised using a hoist mechanism.

X-ray Irradiator (employing a radiation chamber) X-rays are photons and have similar properties to gamma rays emitted by Cobalt-60. X-rays are generated by using an electron beam accelerator (above) and converting the electron beam (up to 7.5 MeV) to photons by accelerating the electrons into a high density material such as tungsten, steel or tantalum. The sudden deceleration of the electrons generates x-rays and waste heat. The creating of the radiation is very similar to an electron beam irradiator (above), including the ability to be turned on and off. The shielding and product conveyance are similar to that of a chamber type gamma irradiator (above). The safety interlocks are similar to both electron beam and chamber type gamma irradiators.

The advantages of x-rays over electron beams are that they have good product penetration (over 24 inches or 60 cm of food product if irradiated on both sides). The advantages of x-rays over both types of gamma irradiators is that they do not require a shielding storage pool. However, there is a substantial loss of energy during the conversion process. Thus, it suffers a severe cost disadvantage when compared to other types of irradiators for the same product volume throughput.

Products are usually irradiated after packaging to minimize the risk of recontamination. This further assures the consumer of a safer product. One tremendous advantage of the irradiation process is that it can be performed on the product in its final retail package. The actual process can take place at the food processing facility, usually after packaging, or at an Irradiation Service Centre. Irradiation Service Centres have been irradiating medical devices, household products, and some food products, for decades to control bacteria. In all cases, the process is conducted by qualified, licensed personnel who follow strict regulated procedures.

Facilities

Facilities for irradiating food are similar to those in operation for sterilizing medical equipment and do not resemble nuclear reactors in any way (Figure 3.1). There are no explosives or materials that could cause widespread dissemination of radioactive material. Facilities using cobalt do not use materials from nuclear defence industries. Facilities used, however, must comply with plant and worker safety requirements of the Nuclear Regulatory Commission and the Occupational Safety and Health Administration.

At an irradiation facility, the radiation source (usually the Cobalt-60) is contained in slender pencil- like stainless steel casing about 18 inches long by 3/8 inch diameter. The casings, in turn, are contained in a lead-lined chamber. Packaged food travels in pallets on a conveyor between 6-1/2 foot thick concrete walls into and through a chamber where it is exposed to the radiation source (gamma rays if Cobalt-60 is used). Pallets may be turned to allow uniform exposure over the route. Radiation dosage is controlled by a computerized rate of passage (conveyor speed) through the chamber.

Doses and Effects of Radiation

In food irradiation, the radiation "dose" that a food receives should not be confused with something added to the food. The "dose" for food irradiation is the amount of radiation absorbed by the food

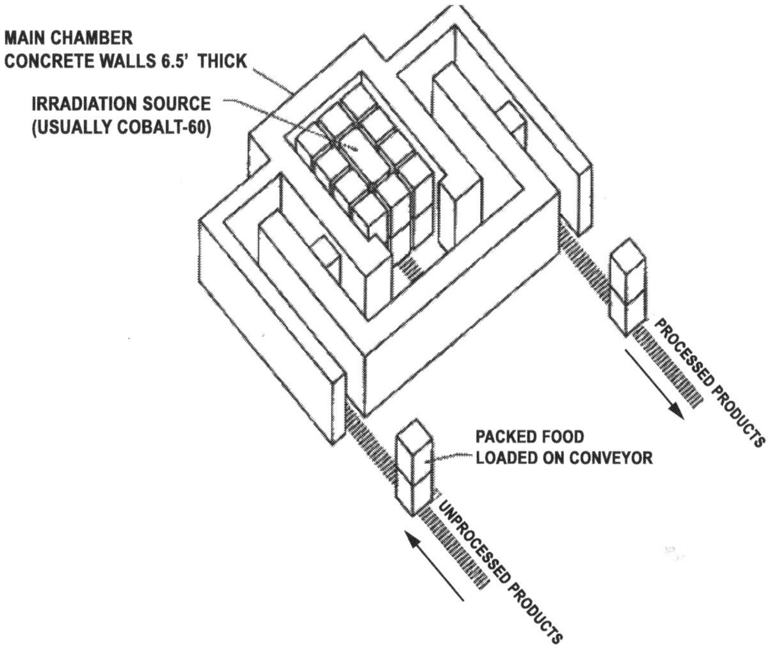

Figure 3.1 ***Typical Irradiation Facility***
(Source: Irradiation Technology Inc;
Rockaway NJ; Adapted FDA Consumer; July –August 1986 : 14-15)

and it is not the same as the level of energy transmitted from the radiation sources. The dose is controlled by the intensity of radiation and the length of time the food is exposed.

Terms used to describe this "dose" or amount of absorbed radiation, are unfamiliar and confusing to most people. In the past, the term used was rad, short for "radiation absorbed dose," which is 100 ergs absorbed by 1 gram of matter. The rad has been replaced by the gray (Gy). One gray is equal to 100 rads or 0.00024 Calorie (i.e., diet calorie) per kilogram of food. (0.00024 Calorie per kilogram equals 0.0001 Calorie per pound.)

The FDA's regulations describe radiation levels in terms of the kilogray (kGy), equal to 1000 Gy. The dose (number of kGy) permitted varies according to the type of food and the desired action.

Treatment levels have been approved by FDA as follows:

"Low" doses (up to 1 kGy) designed to:

- *control insects in grains.*
- *inhibit sprouting in white potatoes.*
- *control trichinae in pork.*
- *inhibit decay and control insects in fruits and vegetables.*

"Medium" doses (1–10 kGy) designed to:

- *control Salmonella, Shigella, Campylobacter, Yersinia, and E. coli in meat, poultry and fish.*
- *delay mould growth on strawberries and other fruits.*

"High" doses (greater than 10 kGy) designed to:

- *kill microorganisms and insects in spices.*
- *commercially sterilize foods, destroying all microorganisms of public health concern (e.g., to sterilize food to the same degree as if they were thermally sterilized (canned); however, no commercial applications of this dosage have been approved by FDA for use in the US with the exception of some special hospital diets for immune-deficient patients.)*

When radiation energy is absorbed by food, it causes a variety of chemical and physical reactions. The amount of energy the food absorbs is controlled so the changes produced have desirable food preservation effects while maintaining the safety, quality, and wholesomeness of the food. The food itself does not become radioactive.

Perishable foods treated with all but high doses, however, must still be refrigerated. Refrigerated storage life is extended, but the need for cold storage is not replaced by irradiation. Microorganisms are destroyed by radiation more easily than enzymes which lead to food deterioration (colour, flavour, and texture changes, for example). Many enzymes survive the current levels of radiation processing, although refrigeration can slow down many enzymatic changes. In addition, not all microorganisms are destroyed. The process does not protect the food against reinfestation or contamination. Therefore, irradiated perishable foods are still considered perishable. In Table 3-1 is shown different sources of radiation and its characteristics.

Table 3-1 ***Different Sources of Radiation***

(Source: Karel, M., and Lund, D. B. Physical Principles of Food Preservation, 2nd. ed., Chapter 11)

Radiation	Nature of Rays	Typical Energy in MeV (1 eV = 10^{-19} J)	Penetration
X-rays	Man made photons	0.01–10	20–150
Gamma Rays	Photons from isotopes	0.1–Several	30-100
Beta Rays	Electrons from isotopes	0.01–1	0.1–0.5
Cathode Rays	Electrons from accelerators	0.1–10	0.1–5
Protons	Protons from accelerators	MeV range	0.003
Neutrons	Neutrons from fission or isotopes	MeV range	10
Alpha Rays	He nuclei from isotopes	1–10	0.005

Effects of Irradiation on Microbial Inactivation

Irradiation is currently used as a tool to control naturally occurring processes such as ripening or senescence of raw fruits and vegetables and is an effective way to inactivate spoilage and pathogenic microorganisms. Irradiation causes microbial death by inhibiting DNA synthesis.

Other mechanisms involved in irradiation microbial inactivation are cell membrane alteration, denaturation of enzymes, alterations in ribonucleic acid (RNA) synthesis, effects on phosphorylation, and DNA compositional changes. According to the dose used and the goal of the treatment, food irradiation can be classified into three categories:

- **Radurization** *is a process comparable with thermal pasteurization. The goal of radurization is to reduce the number of spoilage microorganisms, using doses generally below 10 kGy.*

- **Radicidation** *is a process in which the irradiation dose is enough to reduce specific non-spore-forming microbial pathogens. Doses generally range from 2.5 to 10 kGy, depending on the food being treated.*

- **Radappertization** *is a process designed to inactivate spore-forming pathogenic bacteria, similar to thermal sterilization. Irradiation doses must be between 10 and 50 kGy.*

Moulds and Gram-positive vegetative bacteria are more tolerant than Gram-negative bacteria.

Resistance of *Salmonella typhimurium* increased at reduced irradiation temperatures. Chemical compounds with nutritional or flavour functions can also be affected by ionizing irradiation. At higher irradiation doses, such as those required for food sterilization, some vitamins such as A, B_1, C, E, and K can degrade to some extent.

Irradiation may cause some changes in the sensory characteristics of food and the functional properties of food components. Irradiation initiates the autoxidation of fats, which gives rise to rancid off flavours. The extent of irradiation-induced lipid oxidation depends on factors associated with oxidation such as temperature, oxygen availability, fat composition, and pro-oxidants.

Worldwide Use of Irradiation

Today, food is irradiated in many countries around the world. In France and the Netherlands, large quantities of seafood, vegetables, fish, and frogs' legs are irradiated. Other countries actively involved in food irradiation include Brazil, Mexico, Japan, Belgium, and Israel. In the United States, irradiated hamburger patties are sold in every state. Papayas are irradiated in Hawaii and imported to the mainland. The International Atomic Energy Agency estimated that in 1995 more than 500,000 metric tonnes of food was irradiated throughout the world.

A substantial portion of the spices used in Canada are irradiated. Health Canada is studying requests to apply the process to poultry and red meat.

Labelling Irradiated Foods

India and the Asian Countries

Irradiated food cannot be recognized by sight, smell, taste or touch. Codex Alimentarius Commission has endorsed a green irradiation logo. As per the PFA (Fifth Amendment) Rules, 1994, all packages of irradiated foods to be marketed will be labelled with this logo, along with the words "Processed by Irradiation Method", and the date of irradiation, licence number of the facility and the purpose of irradiation, (Figure 3.2). Consumers will have a free choice to buy radiation processed or non-radiation processed commodity.

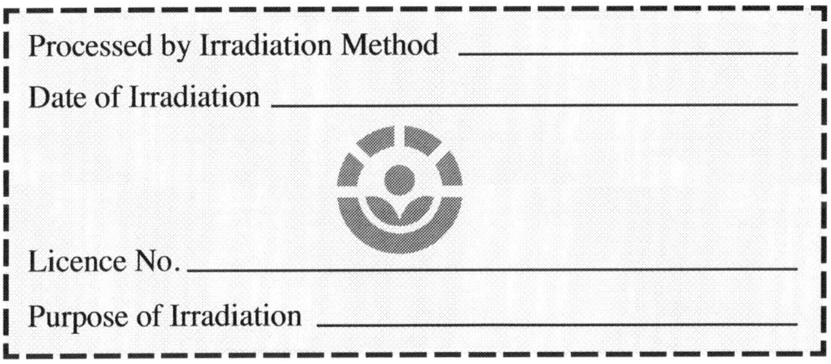

Figure 3.2 ***Label for Processed by Irradiation Method Foods***

The United States of America

Because the beneficial effects of irradiating foods are not discernable to consumers, labelling has been considered essential if consumers are to be adequately informed. After much debate, the FDA approved initial labelling requirements in April, 1986. The regulations required that all produce be labelled at the packing/wholesale and retail levels. Suggested wording and an international symbol (logo) were provided (Figure 3.3).

Figure 3.3 *International Logo for Processed by Irradiation Method Foods.*

At the retail level, the label was required to bear the symbol plus one of these statements: "treated with radiation" or "treated by irradiation." In addition, the manufacturer was allowed to add a phrase which truthfully described the primary purpose of the treatment, such as "treated with radiation to control spoilage."

For unpackaged fruits and vegetables, labels could be on each piece, on the shipping container placed in prominent consumer view, or on a sign near the commodities identifying the use of the treatment.

For irradiated foods sold at the wholesale level, the logo and the wording were still required. In addition, however, they were to be accompanied by the caution "do not irradiate again." Irradiated spices sold in the US do not need to carry the logo if they are in packaged foods where they constitute one small ingredient.

Most packaged spices in stores are usually treated by chemical fumigation, not by irradiation. New legislation in November 1997 (The FDA Modernization Act of 1997, Public Law 105-115) contained food provisions that direct FDA to review its labelling rule for irradiated foods. The agency was told to revise the rule so the disclosure statement is not more prominent than the declaration of ingredients. The ruling does not, however, exclude making a disclosure of irradiation through use of the radura.

References

Dr. Karwe, Mukund V. *Non-Thermal Food Processing.* Department of Food Science; SEBS, Rutgers University. Print.

Bruhn, C. M.; Schutz, H. G.; and Sommer, R. "Attitude change toward food irradiation among conventional and alternative consumers". *Food Technology* 40 (12), (1986) : 86-91. Print.

CAST. "Ionizing energy in food processing and pest control: II. Applications." *Task Force Report No. 115.* Council for Agricultural Science and Technology, Ames, IA. (1989) : 98. Print.

Crawford, L. M. *Statement, News Division, USDA.* May 2, 1990. Print.

International Atomic Energy Agency. *Facts About Food Irradiation.* Vienna, Austria. IAEA/PI/A33E, 1991. Print.

FDA. "Irradiation in the production, processing and handling of food. Federal Register 62 (232) (1997) : 64107-64121. Print.

FDA. "Irradiation in the production, processing, and handling of food; final rule". 21 CFR Part 179. Federal Register 51(75), (1986) :13376-13399. Print.

FDA. HHS News Release P 90-27. US Dept. of HHS. Rockville, MD. May 1, 1990.

Food Irradiation. A Backgrounder of the Institute of Food Technologists. December 1997. IFT, Chicago, IL. Print.

Huang, C. L.; Florkowski, A.V.A.; and Hung, Y. *Food irradiation: A revolutionary technology at the crossroads.* University of Georgia, AES Research Report 540, (1987). Print.

IFT. "Perspective on food irradiation". *Food Technology* 41(2), (1987) :100-101. Print.

Kendall, P. "Food irradiation-promises and controversies. Critical Issues". *Food & Nutrition* 3(6). Colorado State Univ. Cooperative Extension Service, Fort Collins, CO. (1985) : 7. Print.

Lecos, C. "The growing use of irradiation to preserve food". *FDA Consumer* 20(6), (1986) :12-15. Print.

Marcotte, M. "Consumer acceptance of irradiation to preserve food". FDA Consumer 20(6), (1988 :12-15. Print.

Matthews, R. "Personal Conversations". Department of Food Science and Nutrition, IFAS, University of Florida, Gainesville, FL. June, 1990.

Olson, D. "Irradiation of Food. Scientific Status Summary". *Food Technology* 52(1), (1998) : 56-62. Print.

Rogan, A. and Glaros, G. *Food irradiation: The process and implications for dietitians.* (1988). Print.

UFFVA. Food irradiation for the produce industry. Special Report. (1986). Print.

Weise Research Associates, Inc. Consumer reaction to the irradiation concept. US Department of Energy By products Utilization Program. Omaha, NE. (1984). Print.

www.123foodscience.com

http://www.iaea.org/worldatom/inforesource/other/food/

http://www.cdc.gov/ncidod/dbmd/diseaseinfo/foodirradiation.htm

http://www.food-irradiation.com/ http://www.fda.gov/

Chapter 4

Pulsed Light Processing Technology

Definition, Description, and Applications

Definition

High intensity pulsed electric field (PEF) processing involves the application of pulses of high voltage (typically 20 - 80 kV/cm) to foods placed between 2 electrodes. PEF treatment is conducted at ambient, sub-ambient, or slightly above ambient temperature for less than 1 s, and energy loss due to heating of foods is minimized. For food quality attributes, PEF technology is considered superior to traditional heat treatment of foods because it avoids or greatly reduces the detrimental changes of the sensory and physical properties of foods.

Although some studies have concluded that PEF preserves the nutritional components of the food, effects of PEF on the chemical and nutritional aspects of foods must be better understood before it is used in food processing. Some important aspects in pulsed electric field technology are the generation of high electric field intensities, the design of chambers that impart uniform treatment to foods with minimum increase in temperature, and the design of electrodes that minimize the effect of electrolysis. The large field intensities are achieved through storing a large amount of energy in a capacitor bank (a series of capacitors) from a DC power supply, which is then discharged in the form of high voltage pulses. Studies on energy requirements have concluded that PEF is an energy-efficient process compared to thermal pasteurization, particularly when a continuous system is used.

Main Concept

Pulsed light systems consist of several common components (Figure 4.1). A high voltage power supply (1) provides electrical power to the storage capacitor (2), which stores electrical energy for the flash lamp (4). The pulse-forming network (3) determines the pulse shape and spectrum characteristics. A trigger signal (5) initiates discharging of the electrical energy to the flash lamp, which is the key element of a pulsed light unit. The flash lamp is filled with an inert gas, such as xenon, which emits broadband radiation that ranges from UV to NIR.

A high-voltage, high-current electrical pulse is applied to the inert gas in the lamp, and the strong collision between electrons and gas molecules cause excitation of the latter, which then emit an intense, very short light pulse (1 μs to 0.1 s).

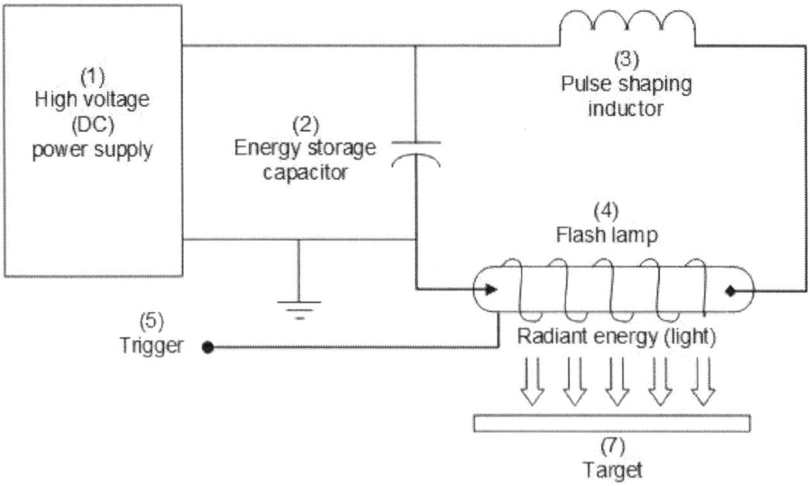

Figure 4.1 ***Functional Diagram of a Pulsed Light System***
(Adapted from Xenon Corporation)

The exact mechanisms by which pulsed light causes cell death are not yet fully understood, but it is generally accepted that UV plays a critical role in microbial inactivation. The antimicrobial effects of UV light on bacteria are attributed to structural changes in the DNA, as well as abnormal ion flow, increased cell membrane permeability and depolarisation of the cell membrane. Some studies also indicated observable injurious effects on yeast cells and mould spores following exposure to pulsed light.

Description of Pulsed Waveforms

PEF may be applied in the form of exponentially decaying, square wave, bipolar, or oscillatory pulses. An exponential decay voltage wave is a unidirectional voltage that rises rapidly to a maximum value and decays slowly to zero.

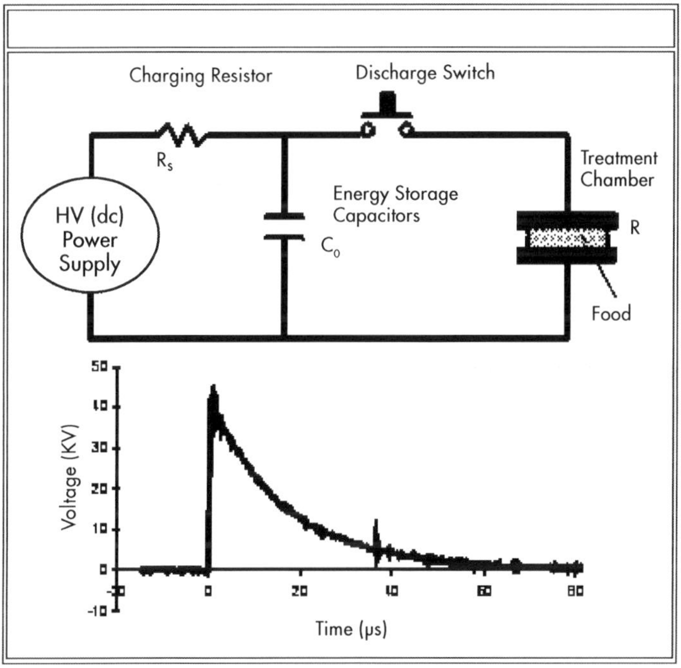

Figure 4.2 ***Electrical Circuit for the Production of Exponential Decay Waveforms***

The circuit in Figure 4.2 may be used to generate an exponential decay waveform. A DC power supply charges a capacitor bank connected in series with a charging resistor (Rs). When a trigger signal is applied, the charge stored in the capacitor flows though the food in the treatment chamber.

Square pulse waveforms are more lethal and more energy efficient than exponential decaying pulses. A square waveform can be obtained by using a pulse-forming network (PFN) consisting of an array of capacitors and inductors and solid state switching devices (Figure 4.3).

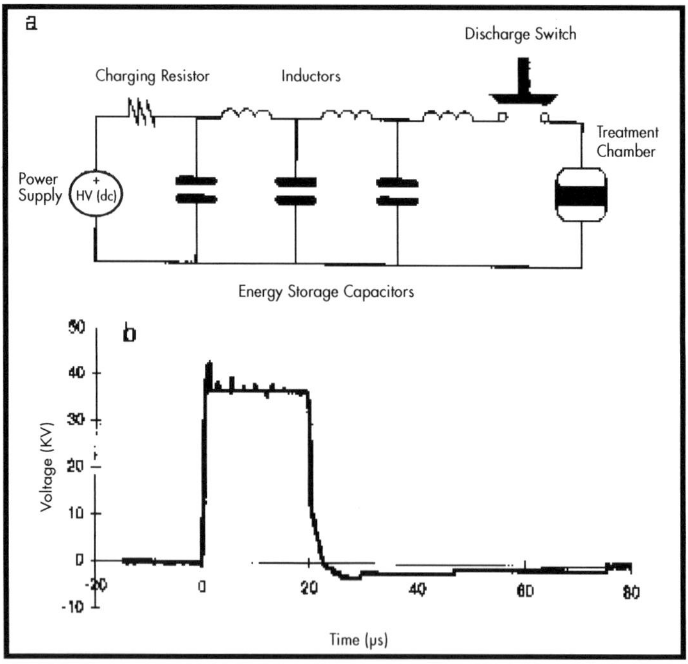

Figure 4.3 ***Square Pulse Generator***
(Using a pulse-forming network of 3 capacitors inductor units and a voltage trace across the treatment chamber.)

The instant-charge-reversal pulses are characterized by a +ve part and -ve part (Fig.4. 4) with various widths and peak field strengths. An instant-charge-reversal pulse width with charge-reversal at the end of the pulse is considerably different from a standard bipolar pulse. In the latter, the polarity of the pulses is reversed alternately with relaxation time between pulses.

Even with a high frequency pulser (for example, 1000 Hz), the dielectric relaxation time at zero voltage between 4 μs square wave pulses is 0.996 ms (Quass 1997). Instant-charge-reversal pulses can drastically reduce energy requirements to as low as 1.3 J/mL (EPRI 1998).

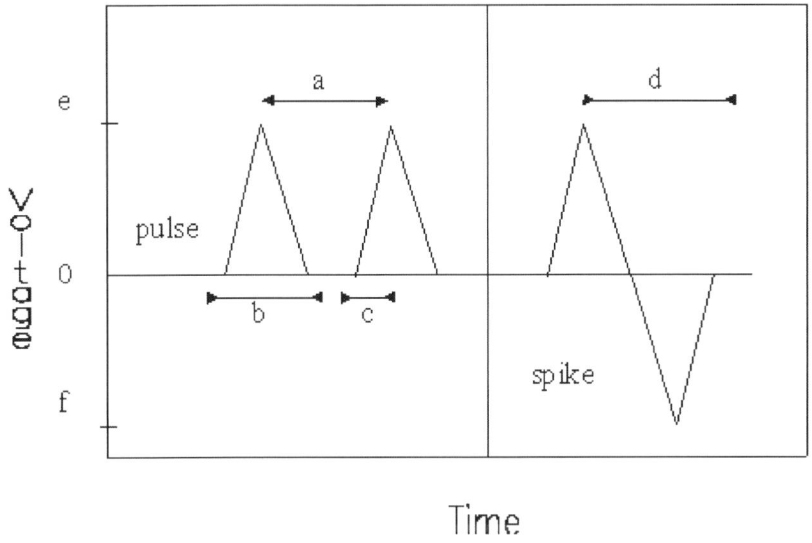

Figure 4.4 ***A Voltage (V) Trace of an Instant-Charge-Reversal Pulse***
(Where a is pulse period (s), b is pulse width (μs), c is a pulse rise time(s) to reach e (kV), d is a spike width(s), e is a peak voltage (kV), and f is a spike voltage (kV) (Ho and others 1995).

Oscillatory decay pulses are the least efficient, because they prevent the cell from being continuously exposed to a high intensity electric field for an extended period of time, thus preventing the cell membrane from irreversible breakdown over a large area.

Treatment Chambers and Equipment

Currently, there are only 2 commercial systems available (one by Pure Pulse Technologies, Inc. and one by Thomson-CSF). Different laboratory- and pilot-scale treatment chambers have been designed and used for PEF treatment of foods. They are classified as static (U-shaped polystyrene and glass coil static chambers) or continuous (chambers with ion conductive membrane, chambers with baffles, enhanced electric field treatment chambers, and coaxial chambers). A continuous flow diagram for PEF processing of foods is illustrated in Fig. 4.5. The test apparatus consists of 5 major components: a high-voltage power supply, an energy storage capacitor, a treatment chamber(s), a pump to conduct food though the treatment chamber(s), a cooling device, voltage, current, temperature measurement devices, and a computer to control operations.

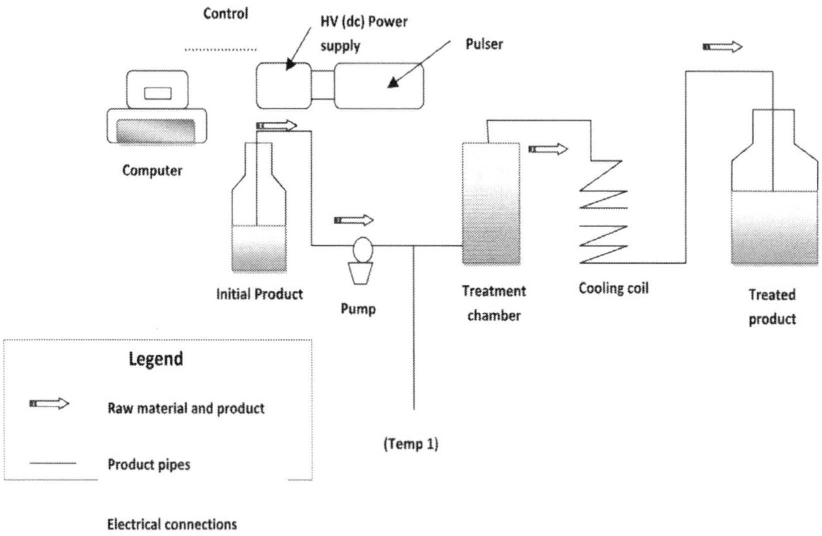

Figure 4.5 ***Continuous PEF Flow Diagram***

Applications of PEF Technology in Food Preservation

To date, PEF has been mainly applied to preserve the quality of foods, such as to improve the shelf-life of bread, milk, orange juice, liquid eggs, and apple juice, and the fermentation properties of brewer's yeast.

Processing of Apple Juice

Simpson and others reported that apple juice from concentrate treated with PEF at 50 kV/cm, 10 pulses, and pulse width of 2 μs and maximum processing temperature of 45 °C had a shelf-life of 28 d compared to a shelf-life of 21 d of fresh-squeezed apple juice. There were no physical or chemical changes in ascorbic acid or sugars in the PEF-treated apple juice and a sensory panel found no significant differences between untreated and electric field treated juices.

Vega Mercado and others reported that PEF extended the shelf-life at 22–25 °C of fresh apple juice and apple juice from concentrate to more than 56 d or 32 d, respectively. There was no apparent change in its physicochemical and sensory properties.

Processing of Orange Juice

Sitzmann (1995) reported on the effectiveness of the ELSTERIL continuous process developed by the food engineers at Krupp Maachinentechnik GmbH in Hamburg, in association with the University of Hamburg, Germany. They reported the reduction of the native microbial flora of freshly squeezed orange juice by 3-log cycles with an applied electric field of 15 kV/cm without significantly affecting its quality.

Zhang and others (1997) evaluated the shelf-life of reconstituted orange juice treated with an integrated PEF pilot plant system. The PEF system consisted of a series of co-field chambers. Temperatures were maintained near ambient with cooling devices between chambers. Three wave shape pulses were used to compare the effectiveness of the processing conditions. Their results confirmed that the square wave is the most effective pulse shape.

In addition, the authors reported that total aerobic counts were reduced by 3- to 4-log cycles under 32 kV/cm. When stored at 4 °C, both heat- and PEF-treated juices had a shelf-life of more than 5 mo. Vitamin C losses were lower and colour was generally better preserved in PEF-treated juices compared to the heat-treated ones up to 90 d (storage temperature of 4 °C or 22 °C) or 15 d (storage temperature of 37 °C) after processing.

Processing of Milk

Dunn and Pearlman conducted a challenge test and shelf-life study with homogenized milk inoculated with *Salmonella Dublin* and treated with 36.7 kV/cm and 40 pulses over a 25-min time period. *Salmonella Dublin* was not detected after PEF treatment or after storage at 7–9 °C for 8 d. The naturally occurring milk bacterial population increased to 107 cfu/mL in the untreated milk, whereas the treated milk showed approximately 4x102 cfu/mL.

Further studies by Dunn indicated less flavour degradation and no chemical or physical changes in milk quality attributes for cheese making. When *Escherichia coli* was used as the challenge bacteria, a 3-log reduction was achieved immediately after the treatment.

Fernandez-Molina and others studied the shelf-life of raw skim milk (0.2% milk fat), treated with PEF at 40 kV/cm, 30 pulses, and treatment time of 2 μs using exponential decaying pulses. The shelf-life of the milk was 2 wk stored at 4 °C; however, treatment of raw skim milk with 80 °C for 6 s followed by PEF treatment at 30 kV/cm, 30 pulses, and pulse width of 2 μs increased the shelf-life up to 22 d, with a total aerobic plate count of 3.6-log cfu/mL and no coliform. The processing temperature did not exceed 28 ° C during PEF treatment of the raw skim milk.

Qin and others (1995b) reported that milk (2% milk fat) subjected to 2 steps of 7 pulses each and 1 step of 6 pulses with an electric field of 40 kV/cm achieved a shelf-life of 2 wk at refrigeration temperature. There was no apparent change in its physical and chemical properties, and no significant differences in sensory attributes between heat pasteurized and PEF treated milk Calderon-Miranda studied the PEF inactivation of *Listeria innocua* suspended in skim milk and its subsequent sensitization to nisin. The microbial population of *L. innocua* was reduced by 2.5-log after PEF treatments at 30, 40 or 50 kV/cm.

The same PEF intensities and subsequent exposure to 10 IU nisin/mL achieved 2-, 2.7- or 3.4-log reduction cycles of *L. innocua*. It appears that there may be an additional inactivation effect as a result of exposure to nisin after PEF. Reina and others studied the inactivation of *Listeria monocytogenes* Scott A in pasteurized whole, 2%, and skim milk with PEF. *Listeria monocytogenes* was reduced 1- to 3-log cycles at 25 °C and 4-log cycles at 50 °C, with no significant differences being found among the 3 milks. The lethal effect of PEF was a function of the field intensity and treatment time.

Processing of Eggs

Some of the earliest studies in egg products were conducted by Dunn and Pearlman (1987) in a static parallel electrode treatment chamber with 2-cm gap using 25 exponentially decaying pulses with peak voltages of around 36 kV. Tests were carried out on liquid eggs, on heat-pasteurized liquid egg products, and on egg products with potassium sorbate and citric acid added as preservatives.

Comparisons were made with regular heat-pasteurized egg products with and without the addition of food preservatives when the eggs were stored at low (4 °C) and high (10 °C) refrigeration temperatures. The study showed the importance of the hurdle approach in shelf-life extension. Its effectiveness was even more evident during storage at low temperatures, where egg products with a final count around 2.7 log cfu/mL stored at 10 °C and 4 °C maintained a low count for 4 and 10 d, respectively, versus a few hours for the heat pasteurized samples.

Other studies on liquid whole eggs (LWE) treated with PEF conducted by Qin and others, and Ma and others showed that PEF treatment decreased the viscosity but increased the colour (in terms of b-carotene concentration) of liquid whole eggs compared to fresh eggs. After sensory panel evaluation with a triangle test, Qin and others (1995b) found no differences between scrambled eggs prepared from fresh eggs and electric field-treated eggs; the latter were preferred over a commercial brand.

In addition to colour analysis of eggs products, Ma and others evaluated the density of fresh and PEF-treated LWE (indicator of egg protein-foaming ability), as well as the strength of sponge cake baked with PEF-treated eggs. The stepwise process used by Ma and others did not cause any difference in density or whiteness between the PEF-treated and fresh LWE. The strength of the sponge cakes prepared with PEF-treated eggs was greater than the cake made with non-processed eggs. This difference in strength was attributed to the lower volume obtained after baking with PEF-treated eggs. The statistical analysis of the sensory evaluation revealed no differences between cakes prepared from PEF processed and fresh LWE.

Processing of Green Pea Soup

Vega-Mercado and others (1996a) exposed pea soup to 2 steps of 16 pulses at 35 kV/cm to prevent an increase in temperature beyond 55 °C during treatment. The shelf-life of the PEF-treated pea soup stored at refrigeration temperature exceeded 4 wk, while 22 or 32 °C were found inappropriate to store the product. There were no apparent changes in the physical and chemical properties or sensory attributes of the pea soup directly after PEF processing or during the 4 wk of storage at refrigeration temperatures.

Current Limitations

Some of the most important current technical drawbacks or limitations of the PEF technology are:

- **The availability of commercial units**, *which is limited to one by Pure Pulse Technologies, Inc., and one by Thomson-CSF. Many pulse-power suppliers are capable of designing and constructing reliable pulsers, but except for these 2 mentioned, the complete PEF systems must be assembled independently. The systems (including treatment chambers and power supply equipments) need to be scaled up to commercial systems.*

- **The presence of bubbles**, *which may lead to non-uniform treatment as well as operational and safety problems. When the* *applied electric field exceeds the dielectric strength of the gas bubbles, partial discharges take place inside the bubbles that can volatize the liquid and therefore increase the volume of the bubbles. The bubbles may become big enough to bridge the gap between the 2 electrodes and may produce a spark. Therefore, air bubbles in the food must be removed, particularly with batch systems. Vacuum degassing or pressurizing the treatment media during processing, using positive back pressure, can minimize the presence of gas. In general, however, the PEF method is not suitable for most of the solid food products containing air bubbles when placed in the treatment chamber.*

- **Limited application,** *which is restricted to food products that can withstand high electric fields. The dielectric property of a food is closely related to its physical structure and chemical composition. Homogeneous liquids with low electrical conductivity provide ideal conditions for continuous treatment with the PEF method. Food products without the addition of salt have conductivity in the range of 0.1 to 0.5 S/m. Products with high electrical conductivity reduce the resistance of the chamber and consequently require more energy to achieve a specific electrical field. Therefore, when processing high salt products, the salt should be added after processing.*

- **The particle size** *of the liquid food in both static and flow treatment modes. The maximum particle size in the liquid must be smaller than the gap of the treatment region in the chamber in order to maintain a proper processing operation.*

- **The lack of methods** *to accurately measure treatment delivery. The number and diversity in equipment, limits the validity of conclusions that can be drawn about the effectiveness of particular process conditions. A method to measure treatment delivery would prevent inconsistent results due to variations in PEF systems. Such a method is not available yet.*

Critical Process Factors and How they Impact Microbial Inactivation

Analysis of Critical Factors

Process Factors

- **Electric field intensity** Electric field intensity is one of the main factors that influences microbial inactivation. The microbial inactivation increases with an increase in the electric field intensity, above the critical transmembrane potential. This is consistent with the electroporation theory, in which the induced potential difference across the cell membrane is proportional to the applied electric field .

Some empirical mathematical models (See Tables 2-4 and 2-5) have been proposed to describe the relationship between the electric field intensity and microbial inactivation. The critical electric field Ec (electric field intensity below which inactivation does not occur) increases with the trans-membrane potential of the cell. Transmembrane potentials, and consequently Ec, are larger for larger cells. Pulse width also influences the critical electric field; for instance, with pulse widths greater than 50 μs, Ec is 4.9 kV/cm. With pulse widths less than 2 μs, Ec is 40 kV/cm. The model of Peleg (Table 5) was used to relate the electric field intensity and applied number of pulses required to inactivate 50% of the cells.

- **Treatment time** Treatment time is defined as the product of the number pulses and the pulse duration. An increase in any of these variables increases microbial inactivation. As noted above, pulse width influences microbial reduction by affecting Ec. Longer widths decrease Ec, which results in higher inactivation; however, an increase in pulse duration may also result in an undesirable food temperature increase.

Optimum processing conditions should therefore be established to obtain the highest inactivation rate with the lowest heating effect.

Hülsheger and others (1981) proposed an inactivation kinetic model that relates microbial survival fraction (S) with PEF treatment time (t). The inactivation of microorganisms increases with an increase in treatment time.

In certain cases, though, the number of pulses increasing inactivation reaches saturation. Such is the case of *Saccharomyces cerevisiae* inactivation by PEF that reaches saturation with 10 pulses of an electric field at 25 kV/cm. Critical treatment time also depends on the electric field intensity applied. Above the critical electric field, critical treatment time decreases with higher electric fields. Barbosa-Cánovas and others reported that for an electric field strength 1.5 times higher than Ec, the critical treatment time would remain constant.

- **Pulse wave shape** Electric field pulses may be applied in the form of exponential decaying, square-wave, oscillatory, bipolar, or instant reverse charges. Oscillatory pulses are the least efficient for microbial inactivation, and square wave pulses are more energy and lethally efficient than exponential decaying pulses.

Bipolar pulses are more lethal than monopolar pulses because a PEF causes movement of charged molecules in the cell membranes of microorganisms, and reversal in the orientation or polarity of the electric field causes a corresponding change in the direction of charged molecules. This difference was reported in *Bacillus* spp. spores and *E. coli*. With bipolar pulses, the alternating changes in the movement of charged molecules cause a stress in the cell membrane and enhance its electric breakdown.

Bipolar pulses also offer the advantages of minimum energy utilization, reduced deposition of solids on the electrode surface, and decreased food electrolysis. As mentioned earlier in this report, the instant-charge-reversal pulse can be described as partially positive at first and partially negative immediately thereafter. This characteristic of the wave shape is influenced by the electrical conductivity of the treated food. In this regard, an increase in conductivity decreases the duration of the positive part of the pulse as well as the span of the negative part, which in turn increases the overall peak/voltage ratio.

The difference between a bipolar and instant charge reverse pulse is the relaxation time between pulses, which is only present in the former. The inactivation effect of an instant-reversal-charge is believed to be due to a significant alternating stress on the microbial cell that causes structural fatigue.

Ho and Mittal reported that instant-reversal-charge may reduce the critical electric field strength required for electroporation of the microbial cell. The effectiveness of this waveform to inactivate micro-organisms compared to other pulse waveforms can save up to 1/5 or 1/6 of total energy and equipment cost. Further work is required to verify the effect of reversal-charge pulses on the inactivation ratio. The inactivation of *Bacillus subtilis* and *Bacillus cereus* spores suspended in NaCl solutions has been reported to be higher when instant reverse pulses and a polarity of electric field chambers with high pulse frequencies are used.

Instant reverse charge has been reported to be effective in inactivation of 5-log cycles of *Bacillus* spp. spores. These researchers established that the survival fraction is not only a function of the temporal pulse area but that even when both bipolar (alternating exponential) and exponential waves had the same area per pulse, bipolar waves yielded a higher inactivation ratio.

A study conducted by Zhang and others showed the effect of square wave, exponentially decaying, and instant-charge-reversal pulses on the shelf-life of orange juice. Three wave shape pulses were used: (a) square waves with peak electric field of 35 kV/cm, an effective pulse width of 37.22 μs, and a pulse rise time of 60 ns; (b) exponential decaying waves with a peak electric field of 62.5 kV/cm, an effective pulse width of 0.57 μs and a pulse rise time of 40 ns; and (c) charge-reversal waves with a peak electric field of 37 kV/cm, an effective pulse width of 0.96 μs, and a pulse rise of 400 ns.

Square wave pulses were more effective, yielding products with longer shelf-lives than those products treated with exponentially decaying and charge reverse pulses. In agreement with this study, Love (et. al.) quantitatively demonstrated the stronger inactivation effect of square wave pulses over other wave shapes.

Qin and others studied the inactivation of *S. cerevisiae* using square and exponential decay waveforms and a peak electric field of 12 kV/cm and 60 J/pulse for both waveforms. The results of this investigation suggested that both waveforms were effective in the microbial inactivation, with square wave pulse waveform being the most effective.

- **Treatment temperature** Experimental results have demonstrated that, in challenge tests, both treatment temperatures and process temperatures impact microbial survival and recovery. PEF treatments at moderate temperatures (~ 50 to 60 °C) have been shown to exhibit synergistic effects on the inactivation of microorganisms. With constant electric field strength, inactivation increases with an increase in temperature. Because the application of electric field intensity does cause some increase in the temperature of the foods, proper cooling is necessary to maintain food temperatures far below those generated by thermal pasteurization.

The effect of temperature was observed when *E. coli* reduction increased from 1 to 6.5-log reduction cycles with a temperature change from 32 to 55 °C. A higher lethal effect of PEF treatment is accomplished by increasing the process temperature to 25 °C, from 5 or 10 °C. This may be due to the increase in the electrical conductivity of the solution at the higher temperature. The authors suggested that the leakage of mobile ions in decoated spores may increase as the temperature is raised due to an increase in average kinetic energy of the ions.

A higher temperature also increases the motion of the solvent molecules in both the surrounding cortex and the core so that the molecules could migrate from one electrode to the other. Additional effects of high treatment temperatures are changes in cell membrane fluidity and permeability, which increases the susceptibility of the to cell to mechanical disruption. Also, a low transmembrane potential decreases Ec and therefore increases inactivation.

Product Factors

- **Conductivity, pH, and ionic strength** The electrical conductivity of a medium (σ, Siems/m), which is defined as the ability to conduct electric current, is an important variable in PEF. Electrical conductivity is the inverse of the resistivity, which is defined by the letter r and measured in ohm-meters (W-m). Foods with large electrical conductivities generate smaller peak electric fields across the treatment chamber and therefore are not feasible for PEF treatment (Barbosa-Cánovas and others 1999).

Inactivation of *Lactobacillus brevis* with PEF showed that as the conductivity of the fluid increased, the resistance of the treatment chamber was reduced (Jayaram and others 1992), which in turn reduced the pulse width and decreased the rate of inactivation. Because an increase in conductivity results from increases the ionic strength of a liquid, an increase in the ionic strength of a food results in a decrease in the inactivation rate. Furthermore, an increase in the difference between the conductivity of a medium and microbial cytoplasm weakens the membrane structure due to an increased flow ionic substance across the membrane. Thus, the inactivation rate of microorganisms increases with decreasing conductivity even with an application of equal input energy (Jayaram and others 1992).

Another study by Dunne and others (1996) with a model system showed resistivity had no effect on PEF effectiveness on *E. coli* and *L. innocua*. These apparent controversial results may be due to the microorganisms or media used. Vega-Mercado and others (1996b) studied the effect of pH and ionic strength of the medium (SMFU) during PEF treatment. The inactivation ratio increases from not detectable to 2.5-log cycles when ionic strength solutions were adjusted from 168 to 28 mM. At 55 kV/cm (8 pulses), as the pH was reduced from 6.8 to 5.7, the inactivation ratio increased from 1.45- to 2.22-log cycles.

The PEF treatment and ionic strength were responsible for electroporation and compression of the cell membrane, whereas the pH of the medium affected the cytoplasm when the electroporation was complete. Dunne and others reported that, depending on the microorganism, acidic pH enhanced microbial inactivation. No mention was made of what microorganisms were affected or what range of pH was used.

- **Particulate foods** Inactivation of microorganisms in liquid-particulate systems has been studied by Dunne et al. *E. coli, L. innocua, Staphyloccocus aureus,* and *Lactobacillus acidophilus* were suspended in a 2 mm diameter alginate beads model, and the effect of variables in PEF microbial inactivation was tested.

The researchers concluded that the process was effective in killing microorganisms embedded in particulates. However, to achieve more than a 5-log cycle reduction, high energy inputs were needed (70–100 J/mL, depending on the bacteria). With those high PEF intensities, the possibility of dielectric breakdown exists- a limitation still to be overcome.

Qin and others reported that dielectric breakdown occurs when air or liquid vapour is present in the food because of the difference in dielectric constant between liquid and gas. Likewise, dielectric breakdown may occur at a particle- to -liquid interface due to differences in electric constants.

- **Hurdle approach** In general, the combination of factors (hurdles) such as pH, ionic strength, and antimicrobial compounds during PEF treatment would be an effective means to aid in the inactivation of microorganisms with PEF.

Microbial Factors

- **Type of microorganisms** Among bacteria, those that are gram-positive are more resistant to PEF than those that are gram-negative. In general, yeasts are more sensitive to electric fields than bacteria due to their larger size, although at low electric fields they seem to be more resistant than gram-negative cells.

A comparison between the inactivation of 2 yeast spp. of different sizes showed that the field intensity needed to achieve the same inactivation level was inversely proportional to cell size. Those results are logical but inconsistent with results by Hülsheger and others. Studies need to continue in this area to better understand the effect of the type of microorganism on microbial inactivation.

- **Concentration of microorganisms** The number of microorganisms in food may have an effect on their inactivation with electric fields. Barbosa-Cánovas and others reported that inactivation of *E. coli* in a model food system of simulated milk ultra filtrate (SMUF) was not affected when the concentration of microorganisms was varied from

103 to 108 cfu/mL after being subjected to 70 kV/cm, 16 pulses, and a pulse width of 2 μs. Increasing the number of *S. cerevisiae* in apple juice resulted in slightly lower inactivation (25 kV/cm, 1 pulse, and pulse width of 25 μs). The effect of microbial concentration on inactivation may be related to cluster formation of yeast cells or possibly concealed microorganisms in low electric field regions.

- **Growth stage of microorganisms** In general, logarithmic phase cells are more sensitive to stress than lag and stationary phase cells. Microbial growth in logarithmic phase is characterized by a high proportion of cells undergoing division, during which the cell membrane is more susceptible to the applied electric field. Hülsheger and others concluded that cells from logarithm growth phase are more sensitive to PEF than from the stationary growth phase. Likewise, *E. coli* cells in the logarithmic phase were more sensitive to PEF treatment when compared to cells in the stationary and lag phases. Studies with *S. cerevisiae* have shown that the susceptibility of actively growing cells to PEF also occurs with yeast cells. For instance, Gaskova and others reported that the killing effect of PEF in the logarithmic phase is 30% greater than of those in stationary phase of growth.

Research Needs

A great deal of research remains to be done before pulsed light technology will be suitable for commercial use. The following is a list of areas where relevant information is lacking:

- *Identification of critical process factors and their effect on microbial inactivation.*

- *Suitability of the technology for solid foods and non-clear liquids where penetration depth is critical.*

- *Potential formation of unpalatable and toxic by-products.*

- *Resistance of common pathogens or surrogate organisms to pulsed light treatments.*

- *Differences between this technology and that of the more conventional UV (254 nm) light treatment.*

- *Mechanisms of microbial inactivation to determine whether they are significantly different from those proposed for UV light.*

- *Understanding of the mechanism and quantification of the benefit attributed to the pulse effect.*

References

Barbosa-Canovas, G. V.; Palou, E., Pothakamury, U. R. and Swanson, B. G. *Application of light pulses in the sterilization of foods and packaging materials. Non-thermal Preservation of Foods.* Chapter 6, (1997) : 139-161. Marcel Dekker. New York. Print.

Chang, J. C. H.; Ossoff, S. F.; Lobe, D. C.; Dorfman, M. H.; Dumais, C. M.; Qualis, R. G.; and Johnson, J. D. "UV inactivation of pathogenic and indicator microorganisms." *Appl Environl Microbiol.* 49 (1985) :1361-1365. Print.

Dunn, J.; Clark, R. W.; Asmus, J. F.; Pearlman, J. S.; Boyer, K.; Pairchaud, F.; and Hofmann, G. "Methods and apparatus for preservation of foodstuffs". US Int. Pat. Appl. No. WO 88/03369, (1988). Print.

Dunn, J.; Clark, R. W.; Asmus, J. F.; Pearlman, J. S.; Boyer, K.; Pairchaud, F.; and Hofmann, G. A. Methods for preservation of foodstuffs. Maxwell Laboratories, Inc. US Patent 5,034,235, (1991). Print.

Dunn, J., Clark, W.; and Ott, T. "Pulsed-light treatment of food and packaging." *Food Technol.* 49(9) (1995) : 95-98. Print.

Dunn, J. "Pulsed light and pulsed electric field for foods and eggs." *Poul Sci.* 75(9), (1996) : 1133-1136. Print.

Dunn, J.E.; Ott, T.M.; Clark, R.W. "Prolongation of shelf-life in perishable pod products." US Patent 5, (1996) : 489,442. Print.

Farkas, J. *Physical methods of food preservation. Food Microbiology. Fundamentals and Frontiers.* M.P. Doyle, L.R. Beauchat, T.J. Montville(eds.). Washington, D.C. ASM Press. (1997) : 497-519. Print.

Jay, J. J. *Modern food microbiology.* Chapman Hall. New York. (1996). Print.

Lagunas-Solar, M. C. "Method of controlling insects and mites with pulsed ultraviolet light." Regents of the University of California. The Regents of the University of California. US Patent 5, (1997) : 607-711. Print.

MacGregor, S. J.; Anderson, J. G.; Fouracre, R. A., Farish, O.; McIlvaney, L.; and Rowan, N. J. "Light Inactivation of food-related pathogenic bacteria using a pulsed power source." *Lett Appl Microbiol.* 27 (1998.) : 67-70. Print.

Chapter 5

Ultrasound Processing Technology

General Overview of the Technology

Ultrasound is a form of energy generated by sound (really pressure) waves of frequencies that are too high to be detected by human ear, i.e. above 16 kHz (Jayasooriya et al. 2004). Ultrasound when propagated through a biological structure, induces compressions and depressions of the medium particles and a high amount of energy can be imparted. In dependence of the frequency used and the sound wave amplitude applied a number of physical, chemical, and biochemical effects can be observed which enables a variety of applications (Knorr et al. 2004).

Ultrasound has been used for a variety of purposes that includes areas as diverse as communication with animals (dog whistles), the detection of flaws in concrete buildings, the synthesis of fine chemicals and the treatment of disease. Despite its wide-ranging uses and exciting developments the study of ultrasound is a young science.

The oldest application, the exploitation of diagnostic ultrasound only dates back to the beginning of the 20th century and ultrasound in processing is even more recent in origin (Mason 2003).

In nature bats and dolphins use low-intensity ultrasound pulses to locate prey; while certain marine animals use high-intensity pulses of ultrasound to stun their victims before capture. In the food industry, a similar division into two distinct categories of ultrasound applications is made (Fellows 2000).

For the classification of ultrasound applications the energy amount of the generated sound field, characterised by sound power (W), sound intensity (W·m-2) or sound energy density (Ws·m-3), is the most important criterion (Knorr et al. 2004). The uses of ultrasound are broadly classified into two groups. Low energy (low power, low-intensity) ultrasound applications involve the use of frequencies higher than 100 kHz at intensities below 1 $W \cdot cm^{-2}$. Low-intensity ultrasound uses a so small power level that the ultrasonic waves cause no physical or chemical alterations in the properties of the material through which the wave passes, that is it is generally non-destructive. They are successfully used for non-invasive monitoring of food processes.

The most widespread application of low- -intensity ultrasound in the food industry is as an analytical technique for providing information about the physicochemical properties of foods, such as composition, structure, and physical state (Fellows 2000, Jayasooriya et al. 2004, Knorr et al. 2004).

Ultrasound has advantages over other traditional analytical techniques because measurements are rapid, non-destructive, and precise, fully automated and might be performed either in a laboratory or on line. One of the most widespread and most promising ultrasonic applications is the utilization of ultrasound for composition measurement. Ultrasonic velocity in fish tissues, chicken, and raw meat mixtures can be related to its composition using semi-empirical equations (Simal et al. 2003).

The other group is high energy (high power, high-intensity) ultrasound which uses intensities higher than 1 $W \cdot cm^{-2}$ (typically in the range 10-1000 $W \cdot cm^{-2}$) at frequencies between 18 and 100 kHz.

Physical, mechanical or chemical effects of ultrasonic waves at this range are capable of altering material properties (e.g. physical disruption, acceleration of certain chemical reactions) (Jayasooriya et al. 2004).

High-intensity ultrasound has been used for many years to generate emulsions, disrupt cells, and disperse aggregated materials. More recently various areas have been identified with greater potential for future development, e.g. modification and control of crystallization processes, degassing of liquid foods, enzymes inactivation, enhanced drying and filtration and the induction of oxidation reactions (Knorr et al. 2004, Zheng and Sun 2006).

The beneficial use of the sound energy is realized through the various effects the ultrasound generates upon the medium where it transmits. Physical, mechanical or chemical effects of ultrasonic waves at this range are capable of altering material properties through generation of immense pressure, shear and temperature gradient in the medium through which they propagate.

During the sonication process, longitudinal waves are created when a sonic wave meets a liquid medium, thereby creating regions of alternating compression and expansion. These regions of pressure change cause cavitation to occur, and gas bubbles are formed in the medium. These bubbles have a larger surface area during the expansion cycle, which increases the diffusion of gas, causing the bubble to expand. A point is reached where the ultrasonic energy provided is not sufficient to retain the vapour phase in the bubble; therefore, rapid condensation occurs. The condensed molecules collide violently, creating shock waves. These shock waves create regions of very high temperature and pressure, reaching up to 5500 °C and 50 MPa.

Cavitation can result in the occurrence of micro streaming which is able to enhance heat and mass transfer (Jayasooriya et al. 2004, Zheng and Sun 2006). The ability of ultrasound to cause cavitation depends on ultrasound characteristics (e.g. frequency, intensity), product properties (e.g. viscosity, surface tension) and ambient conditions (e.g. temperature, pressure) The ultrasound intensity required to cause cavitation increases markedly above about 100 kHz.

Introduction

Definition, Description, and Applications

The textbook definition of ultrasound is energy generated by sound waves of 20,000 or more vibrations per second. Presently, most developments of ultrasonics (sonication) for food applications are non-microbial in nature. High frequencies in the range of 0.1 to 20 MHz, pulsed operation and low power levels (100 mW) are used for non-destructive testing. Ultrasonic excitation is being examined for non-destructive evaluation of the internal quality and latent defects of whole fruits and vegetables in a manner similar to the use of ultrasound for viewing the developing foetus in a mother's womb.

Floros and Liang noted the use of low intensity high-frequency ultrasound for improvement of food product/process monitoring due to the acceleration of diffusion. These industrial applications include texture, viscosity, and concentration measurements of many solid or fluid foods; composition determination of eggs, meats, fruits, and vegetables, dairy, and other products; thickness, flow level, and temperature measurements for monitoring and control of several processes; and non-destructive inspection of egg shells and food packages.

Floros and Liang also listed direct process improvements such as cleaning surfaces, enhancement of dewatering, drying, and filtration, inactivation of microorganisms and enzymes, disruption of cells, degassing of liquids, acceleration of heat transfer, and extraction processes and enhancement of any process dependent upon diffusion. It is evident that ultrasound technology has a wide range of current and future applications in the food industry.

Principal Mechanism of High Power Ultrasound

The fundamental effect of ultrasound on a continuum fluid is to impose an acoustic pressure (P_a) in addition to the hydrostatic pressure already acting on the medium. The acoustic pressure is a sinusoidal wave dependent on time (t), frequency (f), and the maximum pressure amplitude of the wave, $P_{a,\ max}$ (Muthukumaran, Kentish, Stevens, & Ashokkumar, (2006):

$$P_a = P_{a,\ max} \sin(2\pi f t) \qquad \text{Eq. 5.1}$$

The maximum pressure amplitude of the wave (Pa, max) is directly proportional to the power input of the transducer. At low intensity (amplitude), the pressure wave induces motion and mixing within the fluid, so called acoustic streaming.

At higher intensities, the local pressure in the expansion phase of the cycle falls below the vapour pressure of the liquid, causing tiny bubbles to grow (created from existing gas nuclei within the fluid). A further increase generates negative transient pressures within the fluid, enhancing bubble growth and producing new cavities by the tensioning effect on the fluid.

During the compression cycle, the bubble shrinks and their contents are absorbed back into the liquid. However, since the surface area of the bubble is now larger, not all of the vapour is absorbed back into the liquid and thus the bubble grows over a number of cycles. Within a critical size range the oscillation of the bubble wall matches that of the applied frequency of the sound waves causing the bubble to implode during a single compression cycle (Moholkar, Rekveld, & Warmoeskerken, 2000).

This process of compression and rarefaction of the medium particles and the consequent collapse of the bubbles comprises the well-known phenomenon of cavitation, the most important effect in high power ultrasonics. The conditions within these imploding bubbles can be dramatic, with temperatures of 5000 K and pressures of up to 1000 atmospheres, which in turn produces very high shear energy waves and turbulence in the cavitation zone. It is the combination of these factors (heat, pressure, and turbulence) which is used to accelerate mass transfer in chemical reactions, create new reaction pathways, break down and dislodge particles (when cavitation in proximity of a solid surface) or even generate different products from those obtained under conventional conditions. When sound waves reflect on a solid surface or an air-water interface a standing wave can be formed. The acoustic pressure at the nodes is equal to zero, whereas at the anti-node the acoustic pressure fluctuates from a maximum to a minimum.

Leighton and Laborde et al. explain that bubbles smaller than the resonance size accumulate at the anti-node, whereas bubbles larger than the resonance size accumulate at the node and consequently coalesce as they collide. This process of bubble transport and growth at the nodes and anti-nodes is called micro streaming and is the main mechanism for ultrasonic degassing.

Ultrasound (i.e., mechanical waves at a frequency above the threshold of human hearing) can be divided into three frequency ranges:

- *Power ultrasound (16–100 kHz)*
- *High frequency ultrasound (100 kHz–1 MHz)*
- *Diagnostic ultrasound (1–10 MHz)*

The work published by Lorimer & Mason shows that the frequency is inversely proportional to the bubble size. Therefore, low frequency ultrasound (that is, power ultrasound 16–100 kHz) generates large cavitation bubbles resulting in higher temperatures and pressures in the cavitation zone. As the frequency increases the cavitation zone becomes less violent and in the megahertz range no cavitation is observed anymore and the main mechanism is acoustic streaming. While medical imaging operates at frequencies in the megahertz range, most industrial applications (processing of chemicals, food as well as cleaning) operate between 16 and 100 kHz because cavitation can be produced within this frequency range.

The use of ultrasonics in industrial processes has two main requirements; a liquid medium (even if the liquid element forms only 5% of the overall medium) and a source of high-energy vibrations (the ultrasound). The vibrational energy source is called a transducer which transfers the vibration (after amplification) to the so-called sonotrode or probe, which is in direct contact with the processing medium.

There are two main types of transducers: piezoelectric and magnetostrictive. Piezoelectric transducers are the most commonly used in commercial scale applications due to their scalability; i.e., the maximum power per single transducer is generally higher than magnetostrictive transducers.

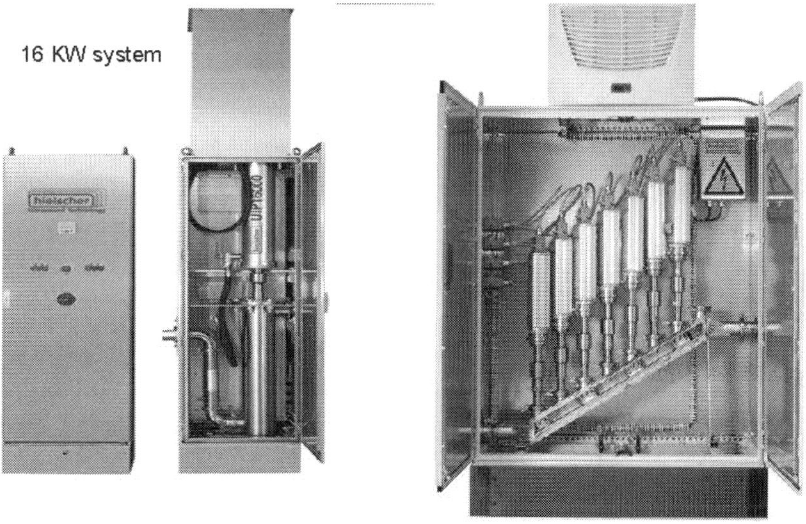

Figure 5.1 ***Industrial Scale System 20 KHz; 16KW***
(Source: CSIRO)

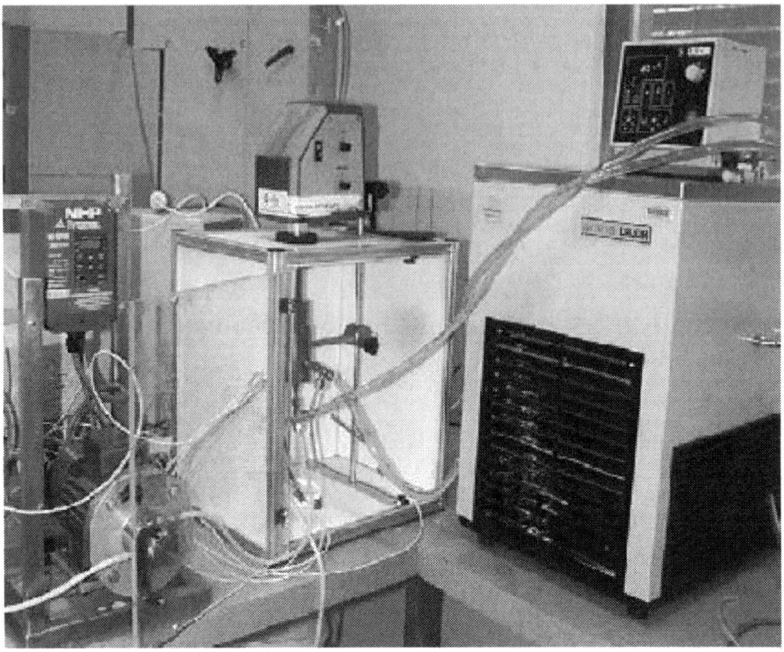

Figure 5.2 ***Ultrasound Equipment***
(Source: Gustavo V. Barbosa-Cánovas Center for Non-thermal Processing of Foods ; Washington State University)

Process Parameters

Energy and Intensity

Ultrasonic liquid processing can be described by the following parameters: amplitude (See previous Section Eq. 5.1), pressure, temperature, viscosity, and concentration of solids. The result or outcome (e.g., % improved extraction yield and/or rate) is a function of:

- **Energy** *The energy input per volume treated material (in kWh/L).*

- **Intensity** *The actual power output per surface area of the sonotrode (in Watts/cm²).*

Where, the energy input is the product of power output (here: kW) and the time of exposure time (here: hours). Both energy and intensity are independent of scale and thus any ultrasonic process will be scaleable using these two parameters (Hielscher, 2005).

Pressure

Increasing the external pressure (as controlled by the back pressure) increases the cavitation threshold and thus the number of cavitation bubbles is reduced (Muthukumaran et al., 2006). On the other hand, increasing the external pressure will increase the pressure in the bubble at the moment of collapse resulting in a more rapid but violent collapse (Lorimer & Mason, 1987). Therefore, increasing the back pressure can be an effective tool in intensifying the process without having to increase the amplitude (Hielscher, 2005).

Temperature and Viscosity

Temperature affects the vapour pressure, surface tension, and viscosity of the liquid medium (Muthukumaran et al., 2006). While increased temperature increases the number of cavitation bubbles, the collapse is 'cushioned' or 'dampened' by the higher vapour pressure. Cavitation bubbles form less easily in a highly viscous environment. Increased temperature decreases the viscosity allowing for a more violent collapse. Thus, there is an optimum temperature at which the viscosity is low enough to form enough violent cavitation bubbles, yet the temperature is low enough to avoid the dampening effect by a high vapour pressure.

It becomes clear that there are many process parameters affecting the process output and thus it takes time and effort to scale and fine-tune the process with the goal to achieve the maximum result with a minimum amount of energy (as that determines the number of transducers required for the commercial application).

Application of Ultrasound in Food Processing

Ultrasound technology can be used in the following processes:

- *Pasteurisation at mild heat*
- *Enhancement of biocides*
- *Extraction*
- *Emulsification*
- *Enzyme activation*
- *Enzyme inactivation*
- *Triggering of biochemical reactions*
- *Crystallisation*
- *Viscosity alteration*
- *Filtration/sieving/ dewatering*
- *Spraying/coating*
- *Encapsulation*
- *Fermentations*
- *Cleaning*
- *Anti-fouling*
- *High shear mixing*
- *Low temperature drying*
- *De-foaming*
- *Waste sludge and effluent treatment*
- *Atomisation*
- *Cutting*
- *Degassing*

Developments in the application of ultrasound in processing began in the years preceding the Second World War when it was being investigated for a range of technologies including emulsification and surface cleaning. By the 1960s the industrial uses of power ultrasound were accepted and being used in cleaning and plastic welding which continue to be major applications (Mason 2003). The possibility of using low-intensity ultrasound to characterize foods was first realized over 60 years ago; however, it is only recently that the full potential of the technique has been realized.

There are a number of reasons for the current interest in ultrasound. The food industry is becoming increasingly aware of the importance of developing new analytical techniques to study complex food materials, and to monitor properties of foods during processing; ultrasonic techniques are ideally suited to both of these applications. Ultrasonic instrumentation can be fully automated and make rapid and precise measurements. Ultrasound is non-destructive and non-invasive, can easily be adapted for on-line applications, and used to analyse systems that are optically opaque.

Within food technology we can find almost all of the examples of processing to which ultrasound can be applied. Until recently the majority of applications of ultrasound in food technology involved non-invasive analysis with particular reference to quality assessment. Such applications use techniques that are similar to those developed in diagnostic medicine, or non-destructive testing, using high frequency low power ultrasound.

Examples of the use of such technologies are found in the location of foreign bodies in food, the analysis of droplet size in emulsions of edible fats and oils and the determination of the extent of crystallization in dispersed emulsion droplets. The relationship between measurable ultrasonic properties of foods (velocity, attenuation coefficient, and impedance) and their physicochemical properties (composition, structure, and physical state) is the basis of the ultrasonic analysis. This relationship can be established either empirically by preparing a calibration curve relating the property of interest to the measured ultrasonic property, or theoretically by using equations describing the propagation of ultrasound through materials. By monitoring the attenuation of an ultrasound pulse has proved possible to determine the degree of homogenisation of fat within milk.

The measurement of ultrasound velocity in conjunction with attenuation can be used to estimate the degree of emulsification in such materials. It is possible to determine factors such as the degree of "creaming" of a sample, i.e. the movement of solid particles/fat droplets to the surface. Such information gives details, for example, of the long term stability of fruit juices and the stability of emulsions such as mayonnaise. The combination of velocity and attenuation measurements shows promise as a method for the analysis of edible fats and oil as well as for the determination of the extent of crystallization and melting in dispersed emulsion droplets.

In recent years food technologists have turned their attention to employment of power ultrasound in processing. Its history can be traced back to 1927 when a paper entitled "The Chemical Effects of High Frequency Sound Waves I. A Preliminary Survey" was published. Physical, mechanical, or chemical effects of ultrasonic waves at this range are capable of altering material properties (e.g., disrupting the physical integrity, acceleration of certain chemical reactions) through generation of immense pressure, shear, and temperature gradient in the medium through which they propagate.

High power ultrasonic applications generally depend on complex vibration induced effects in the propagating media, which produces cavitation in liquids or biological tissue. In addition to cavitation, ultrasound is able to weaken the physical structure of the material or medium, provided the dimensions of the media are similar to those of the ultrasonic wavelength used.

One of the major long-established industrial applications of power ultrasound is for cleaning and it has proved to be an extremely efficient technology. Surface cleaning is applicable to a wide range of disciplines and applications (e.g. sensors, filters, substrates, reactors, catalysers, and heat exchangers). Ultrasound has been shown to be particularly effective for in situ cleaning in conjunction with chemical treatment and offers such advantages as: reduced chemical consumption, reduction of direct worker contact with hazardous cleaning chemicals/substances, enhanced cleaning speed, cleaning consistency—the ultrasonic activity is micro in nature and reaches all areas of complex configurations for uniform cleaning, automatic operation and control savings in energy costs, labour, and floor space (Mason, 2003). Possible applications of power ultrasound in the food industry are very wide ranging.

One of the earliest uses of power ultrasound in processing was in emulsification. Emulsions generated with ultrasound are often more stable than those produced conventionally and often require little, if any, surfactant. Investigations have shown that the use of ultrasound as a processing aid can reduce the production time of yoghurt of up to 40%.

Moreover, sonication reduced the normal dependence of the process on the origin of milk as well as improved both the consistency and the texture of the product. It was also found that fish egg exposure to ultrasound of frequency 1 MHz for 35 min, three times a day resulted in the reduction in hatch time for loach from 72 to 60 hours. Several reports in the literature suggest that ultrasonic treatment of seeds before sowing is an effective method of improving crop yield.

One of the original uses of power ultrasound in biochemistry was to break down biological cell walls to liberate the contents. Subsequently it has been shown that power ultrasound can be used to activate immobilized enzymes by increasing the transport of substrate to the enzyme. As far as enzymes are concerned, ultrasound can also be employed as a method of their inhibition.

Chambers reported that pure pepsin was inactivated by sonication probably as a result of cavitation. By applying ultrasound for over three hours, the original activity of peroxidase, responsible for the development of off-flavours and brown pigments, was progressively reduced by 90%. The use of power ultrasound significantly improves the extraction of organic compounds contained within the body of plants and seeds.

The mechanical effects of ultrasound provide a greater penetration of solvent into cellular materials and improve mass transfer. Additional benefit results from the disruption of biological cell walls to facilitate the release of contents. Combined with this effect is enhanced mass transfer, due to the effects of microstreaming which results in a more efficient method for sugar extraction. The sonication accelerated sugar diffusion and gave the higher level of dry matter content and sugar content in juice (Stasiak 2005).

In some cases sonication increased the efficiency of extraction at lower temperatures producing a purer product in a shorter time. By using of ultrasound extraction of tea solids from leaves was improved by nearly 20%. Authors noticed that the majority of material was extracted in the first 10 minutes of sonication.

Zayas reported that an increased yield of the enzyme rennin from calf stomachs has been achieved by using ultrasound. Moreover, the activity of ultrasonic extract was found to be slightly increased in comparison with normal technology. Power ultrasound has proved to be extremely useful in crystallization processes. It serves a number of roles in the initiation of seeding and subsequent crystal formation and growth (Stasiak and Dolatowski 2007). Ultrasound has also been applied to filtration. As a result, the moisture content of slurry containing 50% water was rapidly reduced to 25%; whereas conventional filtration achieves a limit of only 40%.

Another example of ultrasound application of potentially great commercial importance is acoustic drying. Ultrasonically enhanced drying can be carried out at lower temperatures than the conventional methodology which reduces the probability of oxidation or degradation in the material. By employing ultrasound the heat transfer between a solid heated surface and a liquid is increased by approximately 30–60%. Power ultrasound has proved itself an effective method in assisting food freezing and its benefits are wide-ranged. In addition to its traditional application in accelerating ice nucleation process, it can also be applied to freeze concentration and freeze drying processes in order to control crystal size distribution in the frozen products. If it is applied to the process of freezing fresh foodstuffs, ultrasound can not only increase the freezing rate, but also improve the quality of the frozen products.

Application of power ultrasound can also benefit ice cream manufacture by reducing crystal size, preventing incrustation on freezing surface, etc. (Zheng and Sun 2006). Among other applications are improvements in the extraction of flavourings, filtration, mixing and homogenization, and the precipitation of airborne powders, destruction of foams which cause general difficulties in process control e.g. in fermentation. As a result of continued research interest and development in instrumentation, novel applications such as oxidation of unsaturated oils, aging of alcoholic beverages, hydration of acetylene, decalcification of bone, hydrolysis of esters have been developed.

Ultrasonic Inactivation of Microorganisms

The most common techniques currently used to inactivate microorganisms in food products are conventional thermal pasteurization and sterilization. Thermal processing does kill vegetative microorganisms and some spores; however, its effectiveness is dependent on the treatment temperature and time. However, the magnitude of treatment, time, and process temperature is also proportional to the amount of nutrient loss, development of undesirable flavours and deterioration of functional properties of food products.

High power ultrasound is known to damage or disrupt biological cell walls which will result in the destruction of living cells. Unfortunately very high intensities are needed if ultrasound alone is to be used for permanent sterilization. However, the use of ultrasound coupled with other decontamination techniques, such as pressure, heat or extremes of pH is promising.

Thermosonic (heat plus sonication), manosonic (pressure plus sonication), and manothermosonic (heat plus pressure plus sonication) treatments are likely the best methods to inactivate microbes, as they are more energy—efficient and effective in killing microorganisms. The advantages of ultrasound over heat pasteurization include: the minimizing of flavour loss, greater homogeneity and significant energy savings (Piyasena et al., 2003). A considerable amount of data exists regarding the impact of ultrasound on the inactivation of microorganisms (Piyasena et al., 2003).

The effectiveness of an ultrasound treatment is dependent on the type of bacteria being tested. Other factors are amplitude of the ultrasonic waves, exposure time, volume of food being processed, the composition of food and the treatment temperature. Bactericidal effects of ultrasound were observed while suspended in culture medium. According to Lillard, *Salmonella*e attached to broiler skin were reduced upon sonication in peptone at 20 kHz for 30 min.

Results of research carried out by Dolatowski and Stasiak (2002) proved that ultrasound processing was having a significant influence on microbiological contamination of meat. There are a large number of potential applications of high intensity ultrasound in the food industry. Applications of both high- and low-frequency ultrasound in the food industry have already been shown to have considerable potential for either modifying or characterising the properties of foods. In many instances, techniques based on ultrasound have considerable advantages over existing technologies.

Ultrasound in Meat Technology

The use of ultrasound for predicting fat and muscle content in live cattle has been around since the early 1950's. Today, ultrasound technology is routinely used by the beef industry for: evaluating seed stock, identifying dates to slaughter cattle, predicting quality, palatability, and cut-ability in carcasses. One of the most important quality attributes affecting consumer satisfaction and positive perception of beef is its tenderness. Inconsistency in beef tenderness has been rated as one of the major problems faced by the meat industry. Despite long term research interest in achieving consistent eating quality of meat, it still remains as an elusive goal in meat science.

Tenderness is influenced by composition, structural organization and the integrity of skeletal muscle (Jayasooriya et al., 2004). Tenderness of meat is determined by two major components of the skeletal muscle: contractile tissue, which is largely the myofibrillar fraction, and connective tissue fraction (Jayasooriya et al., 2007). Traditional ageing relies on endogenous proteases (Koohmaraie, 1994), however it is time consuming and its effectiveness varies between animals.

Meat tenderness can be controlled by manipulating pre- and post-slaughter conditions through the use of physical methods, such as electrical stimulation (Hwang and Thompson, 2001) and tenderstretch (pelvic suspension) of the prerigor carcass (Fisher et al. 2000).

Postrigor meat tenderness can also be improved by mechanical methods such as blade tenderization, high pressure technology or Hydrodyne process. Chemical and biochemical methods are also being used for tenderization. The importance of tenderness in determining meat acceptability, and the need for processes giving consistent and rapid improvements in tenderness, mean that other processes must be evaluated. One possibility is the use of ultrasound which can cause physical disruption of materials through cavitation related mechanisms such as high shear, pressure and temperature, and formation of free radicals (Jayasooriya et al., 2007).

Applications of ultrasound to provoke changes in physical and chemical properties of meat and meat products have attracted the interest of research workers for the past few decades because it is a pure physical technique, providing an alternative to chemical or thermal means of processing.

Ultrasound was tested for its ability to induce membrane cell disruption that could increase meat tenderness either directly, through the physical weakening of muscle structure, or indirectly, by the activation of proteolysis either by release of cathepsins from lysosomes and/or of Ca^{++} ions from intracellular stores so that it may activate the calpains. Ultrasound treatment of meat has produced inconsistent effects on meat tenderness, with some ultrasound treatments producing no effect on tenderness, while others decreased or increased tenderness (Jayasooriya et al., 2004).

Acoustic parameters (frequency, intensity, duration of treatment, temperature) determine the extent of the desired result achieved from sonication. Some studies show increased tenderness with low frequency ultrasound (22–40 kHz) treatment Zayas and Gorbatow (1978) also reported the improvement of the tenderness of meat immersed in brine, sonifying at frequency of 22 kHz and 1.5–3 $W{\cdot}cm^{-2}$. Experiments carried out on Semimembranosus muscle showed that ultrasound treatment (frequency 25 kHz; intensity 2 $W{\cdot}cm^{-2}$) during rigor mortis period (up to 24 hours post mortem) resulted in improved tenderness of meat during its ageing.

The changes of sarcomere structure as well as higher WHC were observed for sonicated samples (Dolatowski 1999, Dolatowski and Twarda 2004). Alterations in muscle structure, particularly the loss of the typical myofibrillar structure were observed for sonicated horse Semimembranosus muscle pumped with brine. Ultrasound treatment caused fragmentation of myofibrils and disintegration of other cellular components. Ultrasound-assisted process of meat tumbling caused the significant improvement of the yield, tenderness and juiciness of the end product (Dolatowski and Stasiak 1995). Sonication resulted in decreased drip loss and shear force of PSE meat (Dolatowski et al. 2001, Twarda and Dolatowski 2006).

Other studies showed ultrasound did not tenderize meat samples. This may be due to the use of relatively low intensity ultrasound baths (0.29–1.55 $W{\cdot}cm^{-2}$), or high intensity ultrasound (62 $W{\cdot}cm^{-2}$) applied to individual regions of the meat sample for short treatment times (15 s), which may have been insufficient to produce a tenderizing effect. Smith et al. exposed Semitendinosus muscle to ultrasound source at a frequency of 25.9 kHz in a water bath (filled with degassed saline water). Samples that were treated for 2 or 4 min had significantly lower shear force values than the untreated controls. However, the shear force increased after 8 min of treatment or an increased tenderness with short ultrasound treatments (up to 4 min), but decreased tenderness at longer treatment times (8 –16 min). In a study of the effect of high frequency ultrasound on meat texture, pre- and postrigor meat were treated with high frequency, high intensity ultrasound (2.6 MHz, 10 $W \cdot cm^{-2}$). Pre- and post-rigor ultrasound treatments had small effects on raw meat texture, with ultrasound treated meat having a slightly softer raw meat texture after three to six days ageing.

The difference between control and ultrasound treated sample texture had disappeared after 14 days ageing. However, the results of different studies may not be comparable because of differences in the type of muscles used, the animal age, the ultrasonic equipment and efficiency, intensities, frequencies, and the durations of ultrasound treatments. Research in the last decade has shown the potential benefits of ultrasound treatment as an alternative technology for modifying properties of meat and meat products. The meat industry will benefit from further research and development on the applications of ultrasound in modifying physical and/or chemical properties of meat (Jayasooriya et al., 2004).

References

Dolatowski, Zbigniew J.; Stadnik, Joanna; Stasiak, Dariusz. "Applications of ultrasound in food technology." Agricultural University of Lublin. *Acta Sci. Pol., Technol. Aliment.* 6(3) (2007) : 89-99. Print.

Feng, H.; Yang, W.; and Hielscher, T. "Power Ultrasound." DOI: 10.1177/1082013208098814; *Food Science and Technology International* 14 (2008) : 433. Print.

"Application of Ultrasound for improving processing efficiency." Netsanet Shiferaw Terefe; www.csiro.au. Web.

http://www.sagepublications.com. Web.

Chapter 6

Oscillating Magnetic Fields Processing Technology

Definition, Description, and Applications

Static (SMF) and oscillating (OMF) magnetic fields have been explored for their potential as microbial inactivation methods. For SMF, the magnetic field intensity is constant with time, while an OMF is applied in the form of constant amplitude or decaying amplitude sinusoidal waves. The magnetic field may be homogeneous or heterogeneous. In a homogeneous magnetic field, the field intensity B is uniform in the area enclosed by the magnetic field coil, while in a heterogeneous field, B is nonuniform, with the intensities decreasing as distances from the center of the coil increases.

OMF applied in the form of pulses reverses the charge for each pulse, and the intensity of each pulse decreases with time to about 10% of the initial intensity. Preservation of foods with OMF involves sealing food in a plastic bag and subjecting it to 1 to 100 pulses in an OMF with a frequency between 5 to 500 kHz at temperatures in the range of 0 to 50 °C for a total exposure time ranging from 25 to 100 ms.

Frequencies higher than 500 kHz are less effective for microbial inactivation and tend to heat the food material. Magnetic field treatments are carried out at atmospheric pressure and at moderate temperatures. The temperature of the food increases 2–5 °C.

According to Hoffman exposure to magnetic fields causes inhibition in the growth and reproduction of microorganisms. OMF of intensity of 5 to 50 tesla (T) and frequency of 5 to 500 kHz was applied and reduced the number of microorganisms by at least 2-log cycles. Within the magnetic field of 5–50 T, the amount of energy per oscillation coupled to 1 dipole in the DNA is 10–2 to 10–3 eV. OMF of this intensity can be generated using: (1) superconducting coils; (2) coils which produce DC fields; or (3) coils energized by the discharge of energy stored in a capacitor.

Inhibition or stimulation of the growth of microorganisms exposed to magnetic fields may be a result of the magnetic fields themselves or the induced electric fields. The latter is measured in terms of induced electric field strength and induced current density. To differentiate between electric field and magnetic field effects, a cylindrical enclosure containing cells and a medium that can be adapted to in vitro studies employing uniform, single-phase, extremely low frequency (ELF) magnetic fields is recommended.

Inactivation of Microorganisms

Yoshimura classified the effects of magnetic fields on microbial growth and reproduction as (1) inhibitory, (2) stimulatory, and (3) none observable.

Pothakamury and others summarized the effect of magnetic fields on microorganisms as shown in Table 6-1

Table 6-1 ***Effect of Magnetic Fields on Microorganisms***

Microorganism	Type of Magnetic Field	Field Strength (T)	Frequency of pulse (Hz	Effect	Reference
Wine Yeast Cell	Heterogeneous Smagnetic Field	0.04	0	Growth inhibited when exposed for 5, 20, 25, 60, 120, or 150 min; no inhibition for 10, 15, 17 min exposure.	Kimball (1937)
Wine Yeast Cell	Heterogeneous Smagnetic Field	1.1	0	No effect for 5, 10, 20, 40 or 80 min exposure.	Kimball (1937)
Serratia Marcescens	Heterogeneous Smagnetic Field	1.5	-	Growth rate remains same as in controls up to 6 h; growth rate decreases between 6 and 7 h and again increases between 8 and 10 h; at 10 h cell population same as in controls.	Gerenscer and others (1962)
Staphylococcus aureus	Heterogeneous Smagnetic Field	1.5	0	Growth rate increases between 3 and 6 h; then decreases between 6 and 7 h; cell population at 7 h is same as controls.	Gerenscer and others (1962)
Saccharomyces cerevisiae	Heterogeneous Smagnetic Field	0.465	0	Rate of reproduction reduced, incubated for 24, 48 or 72 h.	Van Nostrand and others (1967)
Escherichia coli	Smagnetic Field	0.3	0	Growth simulated	Moore (1979)

Halobacterium halobium, ***Bacillus subtilis***	Smagnetic Field	0.015 0.03 0.06	0	Growth inhibited	Moore [1979]
Pseudomonas aeruginosa, ***Candida albicans***	Omagnetic Field	0.015 0.03 0.06	0.1–0.3	Growth simulated; stimulation increases with increase in frequency.	Moore [1979]
E. coli	Omagnetic Field	0.15	0.05	Inactivation of cells when concentration was 100 cells/mL.	Moore [1979]
***Streptococcus themophilus* in Milk**	Omagnetic field	12.0	6,000 (1 pulse)	Cell population reduced from 25,000 cells/mL to 970.	Moore [1979]
Saccharomyces in Yogurt	Omagnetic field	40.0	416,000 10 (pulses)	Cell population reduced from 3,500 cells/mL to 25.	Hofmann [1985]
***Saccharomyces* in Orange Juice**	Omagnetic field	40.0	416,000 (1 pulse)	Cell population reduced from 25,000 cells/mL to 6.	Hofmann [1985]
Mold spores	Omagnetic field	7.5	8,500 (1 pulse)	Population reduced from 3,000 spores/mL to 1.	Hofmann [1985]
Saccharomyces cerevisiae	Smagnetic field	0.56	0	Decreased growth rate; interaction between temperature and magnetic field only during the logarithmic phase.	Van Nostrand and others [1967]

Hoffman (1985) reported on the inactivation of microorganisms with OMF in milk, yogurt, orange juice, and bread roll dough. According to Hoffman only 1 pulse of OMF was adequate to reduce the bacterial population between 102 and 103 cfu/g. The intensity of the magnetic field required to achieve these effects varied between 2–25 T and a frequency range from 5–500 Hz.

A review of the literature shows that inconsistent results have been obtained on the effect of OMF on microbial growth (Table 1). In some cases OMF stimulated or inhibited microbial growth and, in others, it had no effect on microbial growth. The results presented in Table 6-1 show that, although not well understood, the effect of magnetic fields on the microbial population of foods may depend on the magnetic field intensity, number of pulses, frequency, and property of the food (that is, resistivity, electrical conductivity, and thickness of the foodstuff).

Mechanisms of Microbial Inactivation

SMF or OMF may have some potential to inactivate microorganisms in food. Pothakamury and others reported 2 theories to explain the inactivation mechanisms for cells placed in SMF or OMF. The first theory stated that a "weak" OMF could loosen the bonds between ions and proteins. Many proteins vital to the cell metabolism contain ions. In the presence of a steady background magnetic field such as that of the earth, the biological effects of OMF are more pronounced around particular frequencies, the cyclotron resonance frequency of ions.

An ion entering a magnetic field B at velocity v experiences a force F given by:

$$F = q \vec{v}\ x \qquad \text{(Eq. 6.1)}$$

Figure 6.1 shows the movement of a charged particle in a magnetic field. When v and B are parallel, F is zero. When v is normal to B, the ion moves in a circular path (Figure 6.2). For other orientations between n and B, the ions move in a helical path (Figure 6.3). The frequency at which the ions revolve in the magnetic field is known as the ion's gyrofrequency n, which depends on the charge/mass ratio of the ion and the magnetic field intensity:

$$\mathbf{n = q\,B\,/\,(2\,\pi\,m)} \qquad \text{(Eq. 6.2)}$$

where q is the charge and m is the mass of the ion. Cyclotron resonance occurs when n is equal to the frequency of the magnetic field. At 50 μ T, the resonance frequency of Na^{+} and Ca^{++} is 33.33 and 38.7 Hz, respectively. At cyclotron resonance, energy is transferred selectively from the magnetic field to the ions with n equivalent to frequency of the magnetic field. The interaction site of the magnetic field is the ions in the cell, and they transmit the effects of magnetic fields from the interaction site to other cells, tissues, and organs.

A second theory considers the effect of SMF and OMF on calcium ions bound in calcium-binding proteins, such as calmodulin. The calcium ions continually vibrate about an equilibrium position in the binding site of calmodulin. A steady magnetic field to calmodulin causes the plane of vibration to rotate, or proceed in the direction of magnetic field at a frequency that is exactly equal to the cyclotron frequency of the bounded calcium. Adding a "wobbling" magnetic field at the cyclotron frequency disturbs the precision to such an extent that it loosens the bond between the calcium ion and the calmodulin.

Hoffman suggested that the inactivation of microorganisms may be based on the theory that the OMF may couple energy into the magnetically active parts of large critical molecules such as DNA. Within 5–50 T range, the amount of energy per oscillation coupled to 1 dipole in the DNA is 10–2 to 10–3 eV. Several oscillations and collective assembly of enough local activation may result in the breakdown of covalent bonds in the DNA molecule and inhibition of the growth of microorganisms.

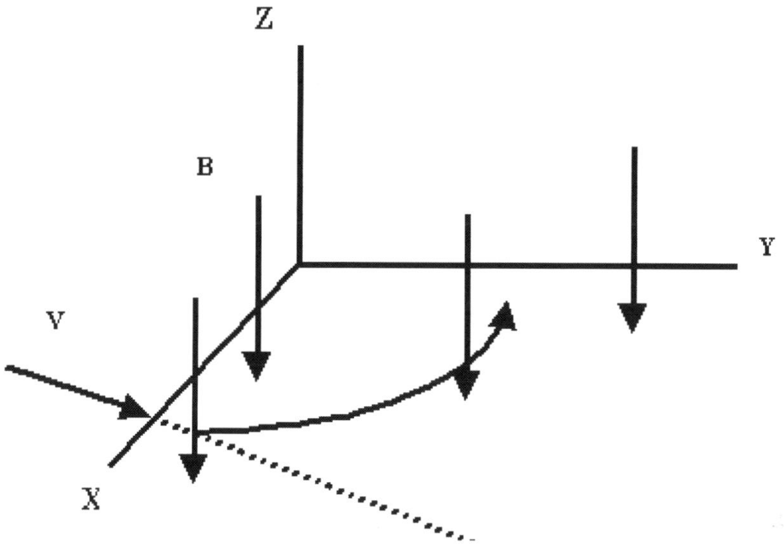

Figure 6.1 ***Charged Particle in a Magnetic Field***

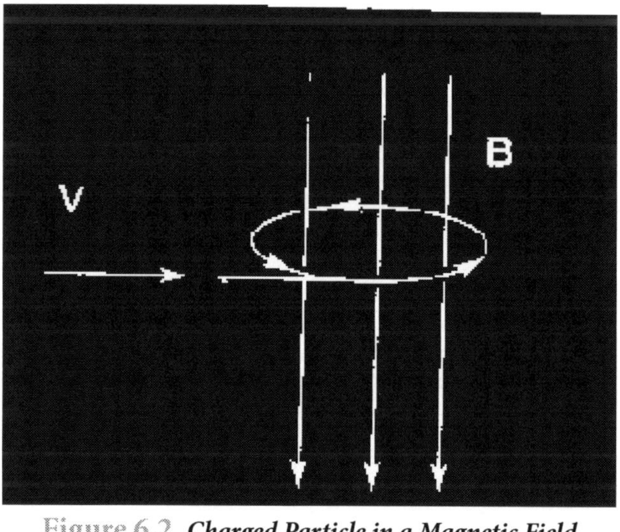

Figure 6.2 ***Charged Particle in a Magnetic Field When V is Normal to B***

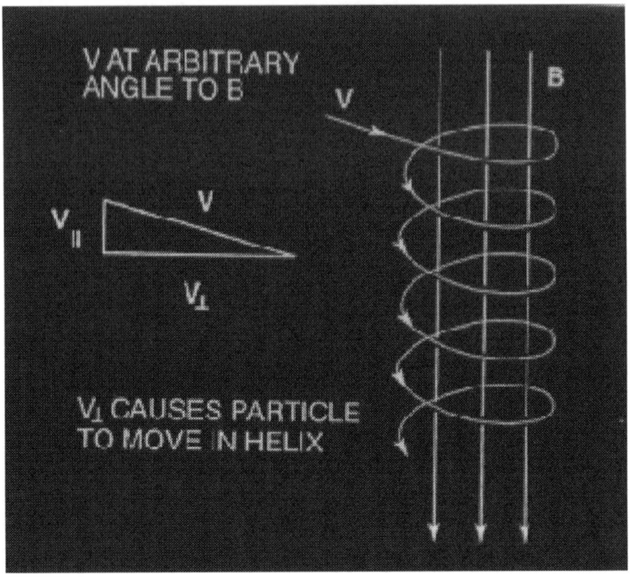

Figure 6.3 ***Charged Particle in a Magnetic Field When V Makes an Arbitrary Angle With B.***

The work of San-Martin and others shows that an externally applied electromagnetic signal at frequencies close to a given resonance and parallel to an SMF (Figure 6.4) may couple to the corresponding ionic species in such a way as to selectively transfer energy to these ions and thus indirectly to the metabolic activities in which they are involved. The earth's total field ranges from 25 to 70 μ T. Most of the slightly and double charged ions of biological interest have corresponding gyrofrequencies in the ELF range 10 to 100 Hz for this field strength.

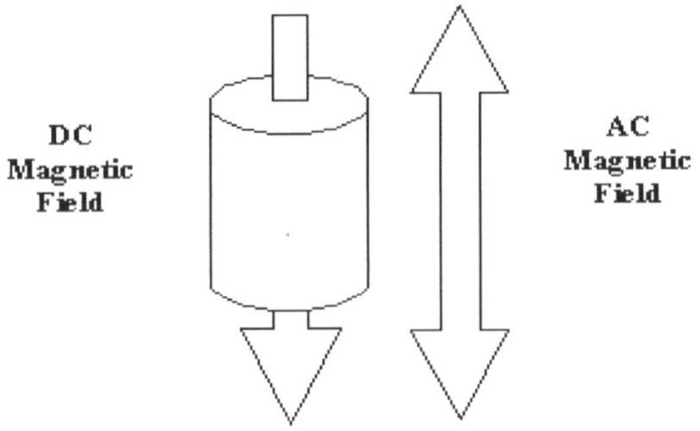

Figure 6.4 ***Required AC and DC Magnetic Field Orientation to Achieve Ion Cyclotron.***

Validation / Critical Process Factors

The critical process factors affecting the inactivation of microbial populations by magnetic fields are not completely understood. Some factors believed to influence microbial inactivation include magnetic field intensity, electrical resistivity, and microbial growth stage.

Magnetic Field

Exposure to a magnetic field may stimulate or inhibit the growth and reproduction of microorganisms. A single pulse of intensity of 5 to 50 T and frequency of 5 to 500 kHz generally reduces the number of microorganisms by at least 2-log cycles. High intensity magnetic fields can affect membrane fluidity and other properties of cells. Inconsistent results of other inactivation studies (See Table 6-1), however, make it impossible to clearly state the microbial inactivation efficiency of magnetic field or to make any predictions about its effects on microbial populations.

Electrical Resistivity

For microorganisms to be inactivated by OMF, foods need to have a high electrical resistivity (greater than 10 to 25 ohms-cm). The applied magnetic field intensity depends on the electrical resistivities and thickness of the food being magnetized, with larger magnetic fields intensities used with products with large resistivity and thickness.

Microbial Growth Stage

Tsuchiya and others, working with homogeneous (7 T) and inhomogeneous (5.2 to 6.1 T and 3.2 to 6.7 T) magnetic fields, found a growth stage dependent response of *Escherichia coli* bacterial cultures. The ratio of cells under magnetic field to cells under geomagnetic field was less than 1 during the first 6 h of treatment and greater than 1 after 24 h. These authors also found that cell survival was greater under inhomogeneous compared with homogeneous fields.

Based on the assumption that magnetic fields could act as a stress factor, cells collected after 30 min of incubation under magnetic field treatment (lag or early lag growth phase) or in the stationary phase after long-term magnetic field treatment were heated to 54 °C. No differences were observed between the treated and control samples. Little else is known about the effect of microbial growth stage on susceptibility to magnetic fields.

Process Deviations

Data acquisition systems must be installed in the processing area to monitor and control the power source, number of pulses, and frequencies applied to the food. Food composition, temperature, size of unit, among other factors also would require control and monitoring to assure constant treatments. Any deviation from the specified conditions such as temperature changes must be continuously recorded and appropriate responses taken. If the system shuts down or fails to deliver the described treatment during processing, the food must be reprocessed to assure quality and safety.

References

Barbosa-Cánovas, G. V.; Pothakamury, U. R.; and Barry, G. S. "State of the art technologies for the stabilization of foods by non-thermal processes: physical methods". In: Barbosa-Cánovas, G.V., and Welti-Chanes, J. (eds.), *Food Preservation by Moisture control.* Lancaster, Technomic Publishing. (1994) : 423-532. Print.

Barbosa-Cánovas, G. V.; Gongora-Nieto, M. M.; and Swanson, B. G. "Non-thermal electrical methods in food preservation." *Food Sci. Int.* 4(5) (1998) : 363-370. Print.

Coughlan, A.; Hall, N. "How magnetic field can influence your ions?" *New Scientist.* 8(4) (1990) :30. Print.

Frankel, R. B.; and Liburdy, R. P. "Biological effects of static magnetic fields". In *Handbook of Biological Effects of Electromagnetic Fields*. Polk, C. and Postow, E. (eds.). 2nd Ed. CRC Press. Boca Raton, FL. Print.

Gerencser, V.F.; Barnothy, M.F.; and Barnothy, J.M. "Inhibition of bacterial growth by magnetic fields." *Nature*, 196 (1962) : 539-541. Print.

Gersdorf, R.; deBoer, F. R.; Wolfrat, J. C.; Muller, F. A.; Roeland, L. W. "The high magnetic facility of the University of Amsterdam, high field magnetism". *Proceedings International Symposium on High Field Magnetism*. Osaka, Japan. (1983) : 277-287. Print.

Hofmann, G. A. "Deactivation of microorganisms by an oscillating magnetic field." US Patent 4,524,079. (1985). Print.

Kimball, G. C. "The growth of yeast on a magnetic fields." J. *Bacteriol.* 35 (1937) : 109-122. Print.

Moore, R. L. "Biological effects of magnetic fields. Studies with microorganisms." *Can. J. Microbiol.*, 25 (1979) :1145-1151. Print.

Pothakamury, U. R.; Barbosa-Cánovas, G. V.; and Swanson, B. G. "Magnetic-field inactivation of microorganisms and generation of biological changes." *Food Technol.* 47(12) (1993) : 85-93. Print.

San-Martin, M. F.; Harte, F. M.; Barbosa-Cánovas, G. V.; and Swanson, B. G. "Magnetic field as a potential non-thermal technology for the inactivation of microorganisms." Washington State University, Biological Systems Engineering, Pullman, WA., USA. (1999) (Unpublished). Print.

Tsuchiya, K.; Nakamura, K.; Okuno, K.; Ano, T. and Shoda, M. "Effect of homogeneous and inhomogeneous high magnetic fields on the growth of *Escherichia coli*." *J Ferment Bioeng* 81(4) (1996) : 343-346. Print.

Van Nostran, F. E.; Reynolds, R.; J. and Hedrick, H .G. "Effects of a high magnetic field at different osmotic pressures and temperatures on multiplication of *Saccharomyces cerevisiae*." *Appl Microbiol.* 15 (1967) : 561-563. Print.

Yoshimura, N. "Application of magnetic action for sterilization of food." *Shokukin Kihatsu* 24(3) (1989) : 46-48. Print.

Chapter 7

Ozone

History of Early Ozone Applications in Food Processing

Some early reported uses of ozone in food processing and preservation have been reported and documented. The examples below represent a sampling of notable cases:

- **Fish and Shellfish Preservation**

Washing of fish with ozonated water extended shelf life for 5 days. Experiments performed by Violle in 1929 found that ozonated seawater spiked with various strains of bacteria (B. typhus, B. coli, etc.) resulted in sterilization that was obtained in fresh water. Further experiments showed that exposure of shellfish to ozonated water did not adversely affect the taste or appearance of the shellfish. Thus, Violle concluded that preozonation of water was suitable treatment for depuration of shellfish. Later work by Salmon and Le Gall built upon the work of Violle and reported that fresh fish, stored under ozonated ice, were edible for between 12 and 16 days. Fish treated with sterilized ice (presumably hypochlorous acid treated) were inedible after the 12th day and, possibly after the 8th day. The original work in this area was carried further by others and reportedly resulted in the *installation of an ozone system designed to sterilize 2000 kg per day in Le Havre and another facility in Boulogne-sur-Mer for treating a daily output of 6000 kg.*

- **Ozone Used in Cold Storage for Meats**

Gaseous ozone was used as a preservation agent in meat and egg storage: Kuprianoff (1953) reported that the first known use of ozone as a food preservation agent was in Cologne, Germany dating back to 1909. Later industrial applications using gas phase ozone for food preservation were reported in 1924 where Hartman stated that "in cold storage ozone is successfully used to prevent the growth of fungi, and eggs have been carried at a relative humidity of 88 and 90 percent and mold developments inhibited with the use of ozone. Hartman summarized by noting that ozone has manifold applications in cold storage and splendid results are being obtained in practice with this reagent every day."

- **Disinfection and Preservation of Fruits and Vegetables**

One of the earliest reported experiments dealing with preservation of fruits was related to ozonation of bananas. Since then, numerous studies have been conducted on a wide variety of fruits and vegetables including carrots, broccoli, pears, peaches, and apples. The vast majority of these studies have reported some degree of shelf life extension and reduction of pathogenic contamination.

Introduction

Ozone, a form of oxygen commonly associated either with its ability to guard against the sun's harmful ultraviolet radiation or with smog, recently gained approval for use in the US food processing industry to help rid food of dangerous pathogens (bacteria, parasites, fungi, and viruses). In July 1997, ozone was deemed "generally recognized as safe" (GRAS) as a disinfectant for foods by an independent panel of experts sponsored by the Electric Power Research Institute.

For any substance commonly used in the US prior to January 1, 1958, the Food and Drug Administration (FDA) allows its use in other products if an independent panel of experts deems the substance and its use as GRAS. The GRAS determination in treating food products was an expansion of uses already approved for ozone.

Ozone has long been recognized as a disinfectant for water, first used in a US water drinking plant in 1940. Today, nearly 200 municipal water treatment plants, from Orlando to Los Angeles, employ ozone to help cleanse their drinking water. Most bottled water is treated with ozone as well, a practice stemming from a 1982 FDA affirmation of ozone as GRAS in this product. Prior to July 1997, however, the only approved use of ozone in food products was for the storage of meat in gaseous ozone, granted by USDA in 1957. Now, processors of fresh fruit, vegetables, poultry, and red meat are examining ozone as one of several new technologies to ensure food safety.

Ozone is a highly reactive form of oxygen, consisting of three oxygen atoms (O_3). It is a potent oxidant/disinfectant that quickly decomposes to diatomic oxygen (O_2), while reacting with targeted organic matter or microorganisms (Figure 7.1). Ozone is naturally generated in the stratosphere, the upper atmospheric layer that protects us from harmful radiation. Gaseous ozone is formed also in the atmosphere during lightning discharges, and on the earth's surface by photochemical reactions, UV sterilization lamps, and high voltage electric arcs.

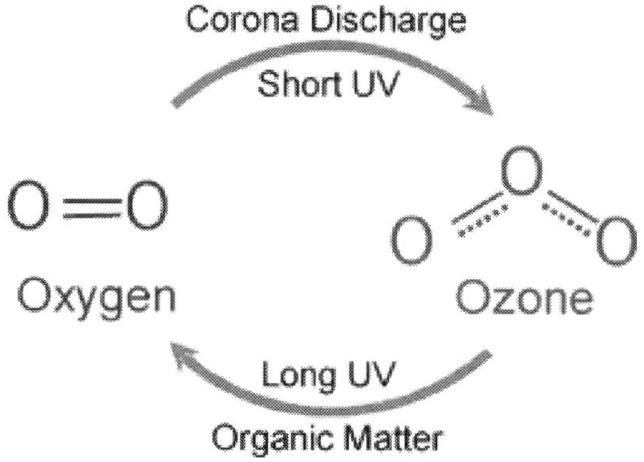

Figure 7.1 ***Ozone Cicle***

Ozone Generation

Corona Discharge Method

This is the most popular type of ozone generator for most industrial and personal uses. While variations of the "hot spark" coronal discharge method of ozone production exist, including medical grade and industrial grade ozone generators, these units usually work by means of a corona discharge tube.

This kind of ozone generators are typically very cost-effective, and do not require an oxygen source other than the ambient air. However, they also produce nitrogen oxides as a by-product. Use of an air dryer can reduce or eliminate nitric acid formation by removing water vapor and increase ozone production. Use of an oxygen concentrator can further increase the ozone production and further reduce the risk of nitric acid formation due to removing not only the water vapor, but also the bulk of the nitrogen.

Ultraviolet Light UV

Ozone generators employ a light source that generates the same narrow-band ultraviolet light that is responsible for the sustenance of the ozone layer in the stratosphere of the Earth. While standard UV ozone generators tend to be less expensive, they usually produce ozone with a concentration of about 2% or lower.

Another disadvantage of this method is that it requires the air to be exposed to the UV source for a longer amount of time, and any air that is not exposed to the UV source will not be treated. This makes UV generators impractical for use in situations that deal with rapidly moving air or water streams (in-duct air sterilization, for example).

Cold Plasma

In the cold plasma method, pure oxygen gas is exposed to a plasma created by dielectric barrier discharge. The diatomic oxygen is split into single atoms, which then recombine in triplets to form ozone. Cold plasma machines utilize pure oxygen as the input source, and produce a maximum concentration of about 5% ozone. They produce far greater quantities of ozone in a given space of time compared to ultraviolet production. However, because cold plasma ozone generators are very expensive, and still require occasional maintenance, they are found less frequently than the previous two types.

The discharges manifest as filamentary transfer of electrons (micro discharges) in a gap between two electrodes. In order to evenly distribute the micro discharges, a dielectric insulator must be used to separate the metallic electrodes and to prevent arcing. Some cold plasma units also have the capability of producing short-lived allotropes of oxygen which include O_4, O_5, O_6, O_7, etc. These species are even more reactive than ordinary O_3.

Ozone Properties

Table 7-1 ***Ozone Properties***
(Source: www.ozoneapplications.com)

Property	Ozone	Oxygen
Molecular Formula	O_3	O_2
Molecular Weight	48	32
Color	Light Blue	Colorless
Smell	- Clothes after being outside on clothesline - Photocopy machines - Smell after lightning storms	Odorless
Solubility in Water (mg/L, 25 °C)	109	8
Density (g/L)	2.144	1.429
Electrochemical Potential (V)	2.07	1.23

Effect of Ozone on Bacteria

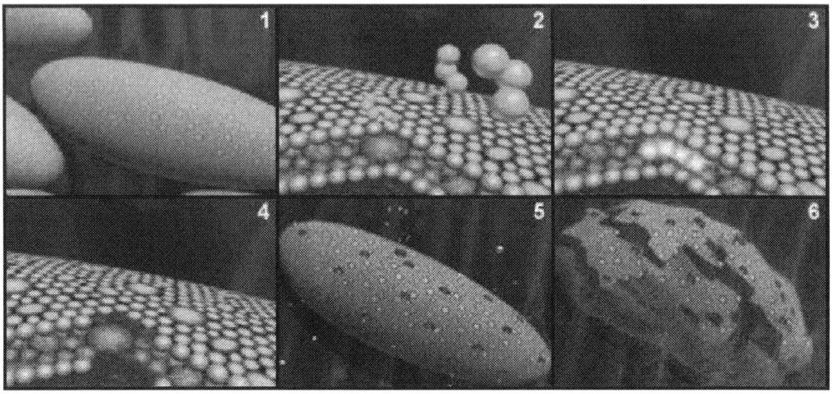

Figure 7.2 ***Illustration of Effect of Ozone on Bacteria***
(Source www.ozoneapplications.com)

1 - Computer generated image of a bacteria cell.
2 - Close-up of ozone molecule coming into contact with bacterial wall.
3 - Ozone penetrating and creating hole in bacterial wall.
4 - Close-up effect of ozone on cell wall.
5 - Bacterial cell after a few ozone molecules come into contact.
6 - Destruction of cell after ozone (cell lysing).

Ozone Effects on Specific Bacteria, Viruses, and Molds

Bacteria are microscopically small, single-cell creatures having a primitive structure. The bacteria body is sealed by a relatively solid-cell membrane. Ozone interferes with the metabolism of bacterium-cells, most likely through inhibiting and blocking the operation of the enzymatic control system. A sufficient amount of ozone breaks through the cell membrane, and this leads to the destruction of the bacteria.

Viruses are small, independent particles, built of crystals and macromolecules. Unlike bacteria, they multiply only within the host cell. They transform protein of the host cell into proteins of their own. Ozone destroys viruses by diffusing through the protein coat into the nucleic acid core, resulting in damage of the viral RNA. At higher concentrations, ozone destroys the capsid, or exterior protein shell by oxidation so DNA (deoxyribonucleic acid), or RNA (ribonucleic acid) structures of the microorganism are affected.

Table 7-2. ***Ozone Effects on Specific Bacteria, Viruses, and Molds.*** *(Source: www.ozoneapplications.com) (1mg/L = 1 ppm)*

Pathogen	Dosage
Aspergillus Niger (Black Mount)	Destroyed by 1.5 to 2 mg/L
Bacillus Bacteria	Destroyed by 0.2 m/L within 30 seconds.
Bacillus Anthracis (causes anthrax in sheep, cattle and pigs. Also a human pathogen)	Ozone susceptible
Bacillus cereus	99% destruction after 5-min at 0.12 mg/L in water.
B. cereus (spores)	99% destruction after 5-min at 2.3 mg/L in water.
Bacillus subtilis	90% reduction at 0.10 ppm for 33 minutes.
Bacteriophage F2	99.99% destruction at 0.41 mg/L for 10-seconds in water.
Botrytis cinerea	3.8 mg/L for 2 minutes
C. difficile	99.999% destruction at 0.6 mg/L for 2 minutes (in water).
Candida Bacteria	Ozone susceptible
Clavibacter michiganense	99.99% destruction at 1.1 mg/L for 5 minutes.
Cladosporium	90% reduction at 0.10 ppm for 12.1 minutes.
Clostridium Bacteria	Ozone susceptible
Clostridium Botulinum Spores. Its toxin paralyses the central nerve system, being a poison multiplying in food and meals.	0.4 to 0.5 mg/L threshold value
Coxsackie Virus A9	95% destruction at 0.035 mg/l for 10-seconds in water.
Coxsackie Virus B5	99.99% destruction at 0.4 mg/l for 2.5-minutes in sludge effluent.

Diphtheria pathogen	Destroyed by 1.5 to 2 mg/L.
Eberth Bacillus (Typhus abdomanalis). Spreads typically by aqueous infection and causes typhoid.	Destroyed by 1.5 to 2 mg/L.
Echo Virus 29 (The virus most sensitive to ozone).	After a contact time of 1 minute at 1 mg/L of ozone, 99.999% killed.
Enteric virus	95% destruction at 4.1 mg/L for 29 minutes in raw wastewater.
Escherichia Coli Bacteria (from feces)	Destroyed by 0.2 mg/L within 30 seconds in air.
E-coli (in clean water)	99.99% destruction at 0.25 mg/L for 1.6 minutes.
E-coli (in wastewater)	99.9% destruction at 2.2 mg/L for 19 minutes.
Encephalomyocarditis Virus	Destroyed to zero level in less than 30 seconds with 0.1 to 0.8 mg/L.
Endamoebic Cysts Bacteria	Ozone susceptible
Enterovirus Virus	Destroyed to zero level in less than 30 seconds with 0.1 to 0.8 mg/L.
Fusarium oxysporum f.sp. lycopersici	1.1 mg/L for 10 minutes
Fusarium oxysporum f.sp. melonogea	99.99 % destruction at 1.1 mg/L for 20 minutes.
GDVII Virus	Destroyed to zero level in less than 30 seconds with 0.1 to 0.8 mg/L.
Hepatitis A Virus	99.5% reduction at 0.25 mg/L for 2-seconds in a phosphate buffer.
Herpes Virus	Destroyed to zero level in less than 30 seconds wit 0.1 to 0.8 mg/L.
Influenza Virus	0.4 to 0.5 mg/L threshold value
Klebs-Loffler Bacillus	Destroyed by 1.5 to 2 mg/L.
Legionella pneumophila	99.99% destruction at 0.32 mg/L for 20 minutes in distilled water.
Mucor piriformis	3.8 mg/L for 2 minutes

Luminescent asidiomycetes (species having no melanin pigment).	Destroyed in 10 minutes at 100 ppm.
Mycobacterium avium	99.9% with a CT value of 0.17 in water (scientifically reviewed document).
Mycobacterium foruitum	90% destruction at 0.25 mg/L for 1.6 minutes in water.
Penicillium Bacteria	Ozone susceptible
Phytophthora parasitica	3.8 mg/L for 2 minutes
Poliomyelitis Virus	99.99% kill with 0.3 to 0.4 mg/L in 3-4 minutes.
Poliovirus type 1	99.5% destruction at 0.25 mg/L for 1.6 minutes in water.
Proteus bacteria	Very susceptible
Pseudomonas Bacteria	Very susceptible
Rhabdovirus Virus	Destroyed to zero level in less than 30 seconds with 0.1 to 0.8 mg/L.
Salmonella Bacteria	Very susceptible
Salmonella typhimurium	99.99% destruction at 0.25 mg/L for 1.67 minutes in water.
Schistosoma Bacteria	Very susceptible
Staph epidermidis	90% reduction at 0.1 ppm for 1.7 min.
Staphylococci	Destroyed by 1.5 to 2.0 mg/L.
Stomatitis Virus	Destroyed to zero level in less than 30 seconds with 0.1 to 0.8 mg/L.
Streptococcus Bacteria	Destroyed by 0.2 mg/L within 30 sec.
Verticillium dahliae	99.99 % destruction at 1.1 mg/L for 20 minutes.
Vesicular Virus	Destroyed to zero level in less than 30 seconds with 0.1 to 0.8 mg/L.
Virbrio Cholera Bacteria	Very susceptible
Vicia Faba progeny	Ozone causes chromosome aberration and its effect is twice that observed by the action of X-rays.

The effect of ozone below a certain critical concentration value is small or zero. Above this level all pathogens are eventually destroyed. This effect is called all-or-none response and the critical level the "threshold value."

Potential Benefits

The strength of the case for using ozone may rest with its versatility and environmental benefits over some existing food sanitizing methods. Ozonated water can be used on food products as a disinfectant wash or spray. When dispersed into water, ozone can kill bacteria—like *E. coli*—faster than traditionally used disinfectants, such as chlorine. Ozone also kills viruses, parasites, and fungi.

The US Environmental Protection Agency, in conjunction with the Safe Drinking Water Act of 1991, confirmed that ozone was effective in ridding water of hazardous pathogens, including chlorine-resistant *Cryptosporidium*. Coupling two processes—washing food with ozonated water and the subsequent ozonation of the recaptured water—reduces the amount of water needed in the food washing system (which lowers costs, particularly for high water users such as fruit and vegetable packers and processors). In addition, any wastewater discharged by an ozonation process used as a substitute for conventional chlorine-based food washing and spraying systems, is free of chlorine residuals, a growing environmental concern in groundwater pollution. Food products treated with ozone are also free of disinfectant residues.

Because it is an unstable gas, ozone decomposes in about 20 minutes into simple oxygen, leaving no trace of the ozone disinfectant on the food. Ozone also acts as a disinfectant in its gaseous state. It can be applied to sanitize food storage rooms and packaging materials, which may help to control insects during storage of foods and prevent spoilage of produce during shipping.

Gaseous ozone is also listed as an alternative disinfectant for water-sensitive produce, such as strawberries and raspberries, in the *Guide to Minimizing Microbial Food Safety Hazards for Fresh Fruits* and Vegetables (a document from FDA and USDA). The Electric Power

Research Institute is examining the use of ozone as a fumigant in food storage beyond the already approved use for meat. Methyl bromide has commonly been used as a fumigant to Gary Lucier

Food Safety Technology: A Potential Role for Ozone?

Ozone is only one of many food sanitizing ingredients and processes being used, examined, or proposed to improve food safety. Chlorine is the most commonly used chemical to kill pathogens on food, but chlorine dioxide, hypochlorite, and trisodium phosphate also have been studied for use in wash water to disinfect food products. Irradiation of meat, through low-dose radiation or electron beams, was approved by the Food and Drug Administration in December 1997.

Steam pasteurization, flash pasteurization, and ultravilet radiation are additional methods that can sanitize food. Each method has its advantages and disadvantages, and research continues on which methods or combinations of sanitizing processes work best for specific foods. prevent insect infestation of commodities such as grapes, raisins, cherries, nuts, and grains, but its use is being phased out under the Clean Air Act (Amendments of 1990). The phase out prohibits US production and importation of methyl bromide starting January 1, 2001. Interestingly, the phase out is intended to halt the depleting effect of methyl bromide on the Earth's protective ozone layer.

Interest in Ozone Systems Builds

The food processing industry has faced mounting concerns in recent years about its ability to provide a consistently safe food supply. Food passes through many hands— from growing, picking, boxing, shipping, to final processing—prior to reaching the consumer. Most past efforts to avoid contamination of food centered on preventing exposure to sewage or animal manure early in the production process. Because of the incidence of food contamination along the entire chain of production, and the recognition that many pathogens—some have recently emerged—are found in even healthy animals, the industry has realized that some form of disinfection, perhaps at multiple points, is necessary.

Each year in the US, an estimated 6.5–33 million illnesses and up to 9,000 deaths are caused by foodborne diseases (AO July 1996). Centuries-old methods of treating food, such as drying, smoking, and use of simple substances like salt, no longer adequately prevent spoilage in today's food marketing system. These methods to prevent contamination can also alter a food's taste. Food processors have turned to other technologies to both decontaminate and preserve products without substantially changing the appearance, taste, texture, or nutrient content of the food.

These methods include steam pasteurization, used principal in meat processing where beef carcasses are exposed to steam for short periods of time; flash pasteurization, a heating process to kill bacteria in juice; and irradiation, which uses low-dose radiation to treat meats, fruits, vegetables, and spices. As a non-thermal method of disinfecting food, ozonation reportedly alters taste little, unlike some heat-based steam and flash pasteurization systems that cook the product. Further, in some foods, ozone proponents indicate flavor is enhanced by ozone's ability to neutralize chemicals, pesticides, and bad tastes from gases produced by ripening or decay.

In 1995, the National Live Stock and Meat Board and various universities conducted research that showed an ozone wash reduced bacterial contamination on beef carcasses to a level equal to conventional carcass trimming and washing methods. (Under specific conditions, hot water washing, an alternative process, resulted in consistently lower bacterial populations on beef carcasses). In mid- May 1998, research was completed by California Polytechnic State University which revealed ozone reduced pathogens on surfaces of lettuce, meat, and poultry.

Now that ozone has received a "generally recognized as safe" designation, a few firms have adopted or begun testing ozone-based systems. Recent televised news reports highlighted a Florida citrus grower washing oranges and grapefruit in ozonated water.

The Vermont Department of Agriculture is examining the potential of ozone to wash apples used in the apple cider industry. Industrial gas-producing companies are developing mechanical systems for processing poultry that filter out biological waste material in poultry chiller water and then add ozone to disinfect the washwater.

In January 1998, two New York-based companies, one an all-natural chicken processor, announced pilot tests of an ozone system in their processing plants. (Before a firm adopts an ozone system, a pilot test is required by the USDA's Food Safety and Inspection Service—FSIS— conducted under FSIS protocols). And in April 1998, an agricultural corporation in California contracted to install an ozone system that is intended to replace a combined chlorine wash and steam pasteurization process.

Recent Studies of Ozonation in Food Processing

In spite of past regulatory restriction limiting usage of ozone in the US food processing industry, numerous studies were undertaken to determine the efficacy and economics of ozonation. The examples below serve to illustrate some of these studies and conclusions.

- **Fruits and Vegetables**

Several controlled studies have been reported in the food science literature relating to the evaluation of treatment of fruits and vegetables with ozonated water. Kondo et al (observed greater than 90% reduction of total bacterial counts upon treatment of Chinese cabbages with ozonated water (2.3 mg/L) for 60 minutes. Treatment of wash water used in processing of carrots has been reported to provide 3 log reduction of bacteria.

Barth et al evaluated ozone exposure for prevention of fungal decay on thornless blackberries. The fruit was harvested and stored for 12 days at 2 °C in 0.0, 0.1 and 0.3 ppm ozone, then evaluated for fungal decay (Botrytis cinerea), anthocyanines, color, and peroxidase activity. Ozone storage suppressed fungal development for 12 days, while 20% of control fruits showed decay. The treated fruit did not show observable injury or defects. Sarig et al showed that ozone at low dosages (0.1 mg/g fruit) for 20 minutes, reduced the levels of fungi yeasts and bacteria on grapes, but that higher doses caused some fruit damage.

- **Poultry**

The poultry processing industry is a large volume consumer of water. The potential for reuse of poultry processing water represents an attractive economic benefit to the industry. In 1996 the USDA approved the use of ozonation for washing of poultry carcasses (provided that the ozone did not come into direct contact with the product). Studies by Chang and Sheldon (1989) reported that a combination of screening, diatomaceous earth filtration and ozonation yielded the highest quality of water with total microbial loads (total coliforms, *E. coli* and *Salmonella* reduced by 99%). A subsequent study, Chang and Sheldon found no significant differences in measures of carcass quality including skin color, taste or shelf life using recycled, ozone treated water as opposed to fresh makeup water. The investigators further reported that a 2.7 log reduction in of total plate count was observed in the recycled water stream. The results of this study showed that, for a typical broiler processing plant (240,000 broilers/day) the savings resulting from chiller water reuse would equate to 50% reduction in discharges and savings of more than $ 45,000 per year.

- **Extension of Food Product Storage Life**

The following are a sampling of studies directed at application of gaseous ozone in food storage facilities. A 1980 study (Gabriel.yants.) showed that cheese stored with periodic ozonation prevented mold growth for 4 months while controls showed mold growth as soon as 1 month. Japanese researchers indicated good results in the treatment of grains, flour and raw noodles with ozone with significant reductions in microbial growth.

Dondo et al. reported that ozone treatment during refrigerated storage stabilized the surface bacteria count on beef and reduced that on fish. Naitoh showed that ozone treatment inside a confectionery factory reduced airborne microorganisms over a 1–1.5 year period, remarkably inhibited bacterial growth, and extended storage life of the product by 7 days.

- **Ozone Treatment of Dye Wastewater**

Several studies have shown ozone and ozone/hydrogen peroxide to be effective in removal of color resulting from dye processing. Dyes used in food processing are typically easier to oxidize with ozone since they are organically based compounds.

Effect of Ozone on Meat

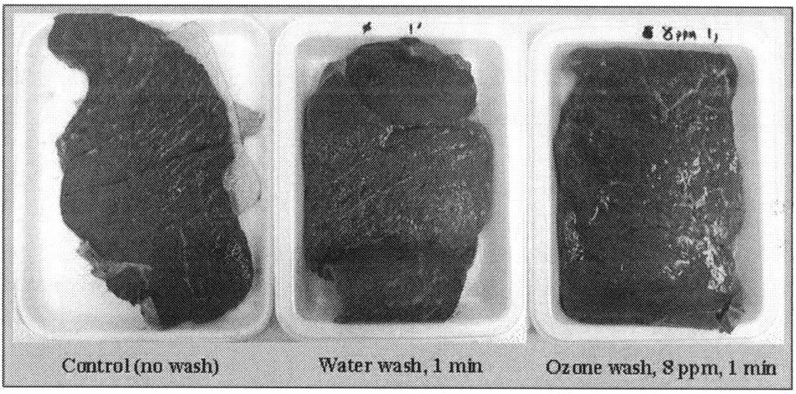

Figure 7.3 ***Illustration of Effect of Ozone on Meat***
(Source: Non-thermal Food Processing; Dr. Mukund V. Karwe; Department of Food Science; SEBS, Rutgers University)

Effect of Ozone on Strawberries

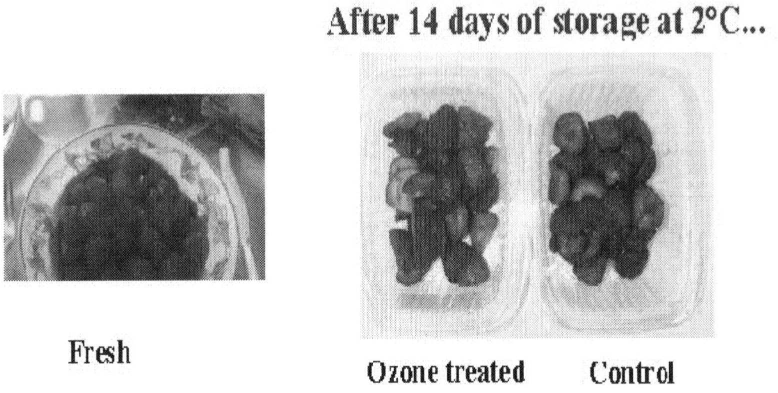

Figure 7.4 ***Illustration of Effect of Ozone on Strawberries***
(Source: Non-thermal Food Processing; Dr. Mukund V. Karwe; Department of Food Science; SEBS, Rutgers University)

Extended Benefits of Ozone

- *Higher levels of ozone can be used before altering the taste or appearance of the produce.*

- *Ozone improves taste appearance over the use of chlorination alone: Better Quality Produce.*

- *Ozone lowers counts of spoilage microorganisms in wash water and on produce surface: Longer Shelf Life.*

- *Ozone keeps wash water cleaner longer: Less Water Usage.*

- *An ozone treatment is capable of destroying pesticides and chemical residues in wash water and on produce.*

- *Eliminate chlorine from a process: No THM's or other chlorinated by-products.*

- *Implementing ozone reduces the risk of cross-contamination of pathogens.*

- *Ozone leaves no chemical residue: No Final Rinse - Less Water Usage.*

- *An ozone system mitigates the need for storage, handling, use, and disposal of chemical sanitation agents.*

- *In some situations, ozone reduces contamination in discharge water: Lower Cost Waste Water Disposal.*

- *Ozone is natural and chemical free allowing the use of ozone in organic food production and processing.*

Equipment

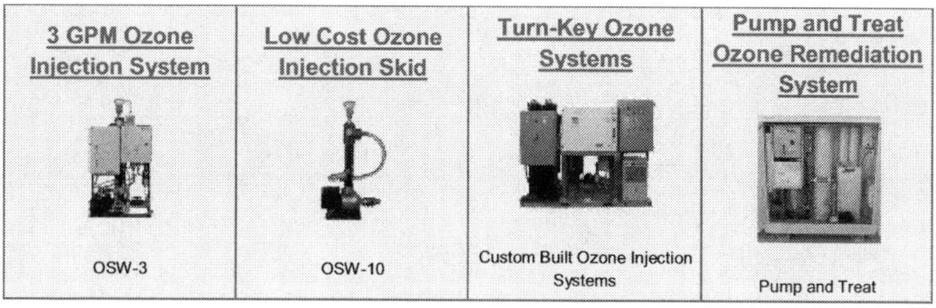

Figure 7.5 ***Illustration of Some Available Ozone Generating Equipment***
(Source: www. ozone solutions.com)

References

Dr. Karwe, Mukund V. "Non-thermal Food Processing." Department of Food Science SEBS, Rutgers University; http://foodsci.rutgers.edu. Web.

"Ozone in Food Processing Applications. Past Experience, Future Potential and Regulatory Issues." Zentox Corporation. Print.

Agricultural Outlook/June-July Economic Research Service/USDA 13 (1998). http://www.econ.ag.gov/Prodsrvs/rept-sag htm#vegetables. Web.

Ramaswamy, Raghupathy; Rodriguez-Romo, Luis; Vurma, Mustafa; Balasubramaniam, V. M. (Bala); Yousef, Ahmed E. *Ozone Technology; Fact Sheet for Food Processors.* Ohio State University. Print.

Khadre, M.A.; Yousef, Ahmed E.; Kim, J. G. "JFS: Concise Reviews in Food Science; Microbiological Aspects of Ozone Applications in Food." A Review. Print.

Smilanick Joseph L.; Wenatchee, W. A. Washington. "Use of ozone in storage and packing facilities." Tree Fruit Postharvest Conference. December 2nd and 3rd, (2003).

Chapter 8

Microwave and Radio Frequency Processing

Definition, Description and Applications

Microwave and radio frequency heating refers to the use of electromagnetic waves of certain frequencies to generate heat in a material . The frequencies allocated by the Federal Communications Commission (FCC) for the purposes of heating are listed in Table 8-1. Typically, microwave food processing uses the 2 frequencies of 2450 and 915 MHz. Of these two, the 2450 MHz frequency is used for home ovens, and both are used in industrial heating. It is worthwhile to note that outside of the United States, frequencies of 433.92, 896 and 2375 MHz are also used.

Radio frequency heating in the United States can be performed at any of the 3 frequencies listed in Table 8-1. As mentioned earlier, there is not much commercial use of these frequencies for food pasteurization or sterilization, although they are used in baking and other processes in the food industry. An overview of food and chemical processing uses of radio frequency can be seen in Kasevich and Minett and Witt.

Table 8-1 ***Frequencies Assigned by the FCC for Industrial, Scientific, and Medical Use***

	Frequency
Radio	13.56 MHz ± 6.68 kHz
	27.12 MHz ± 160.00 kHz
	40.68 MHz ± 20.00 kHz
Microwaves	915 MHz ± 13 MHz
	2450 MHz ± 50 MHz
	5800 MHz ± 75 MHz
	24125 MHz ± 125 MHz

How the Microwaves and Radio Frequencies Generate Heat

Heating with microwave and radio frequency involves primarily 2 mechanisms: dielectric and ionic. Water in the food is often the primary component responsible for dielectric heating. Due to their dipolar nature, water molecules try to follow the electric field associated with electromagnetic radiation as it oscillates at the very high frequencies listed in Table 8-1. Such oscillations of the water molecules produce heat. The second major mechanism of heating with microwaves and radio frequency is through the oscillatory migration of ions in the food that generates heat under the influence of the oscillating electric field.

The rate of heat generation per unit volume, Q, at a particular location in the food during microwave and radio frequency heating can be characterized by the Equation 8.1 (Buffler, 1993; Datta and Anantheswaran 2000):

$$Q = 2\pi f \epsilon_0 \epsilon'' E^2 \qquad \text{Eq. 8.1}$$

where E is the strength of electric field of the wave at that location, f is the frequency (Table 8-1) of the microwaves or the radio frequency waves, ε_0 the permittivity of free space (a physical constant), and ε'' is the dielectric loss factor (a material property called dielectric property) representing the material's ability to absorb the wave.

Not apparent from the above equation, there is another dielectric property called the dielectric constant that affects the strength of the electric field inside the food. The dielectric properties depend on the composition (or formulation) of the food, moisture and salt being the two primary determinants of interest. The subsequent temperature rise in the food depends on the duration of heating, the location in the food, convective heat transfer at the surface, and the extent of evaporation of water inside the food and at its surface.

Advantages of Microwave and Radio Frequency Processing

Microwave and radio frequency heating for pasteurization and sterilization are preferred to the conventional heating for the primary reason that they are rapid and therefore require less time to come up to the desired process temperature. This is particularly true for solid and semi-solid foods that depend on the slow thermal diffusion process in conventional heating. They can approach the benefits of high temperature-short time processing whereby bacterial destruction is achieved, but thermal degradation of the desired components is reduced. This is illustrated in Figure 8.1 for typical time-temperature histories of microwave conventional heated processes.

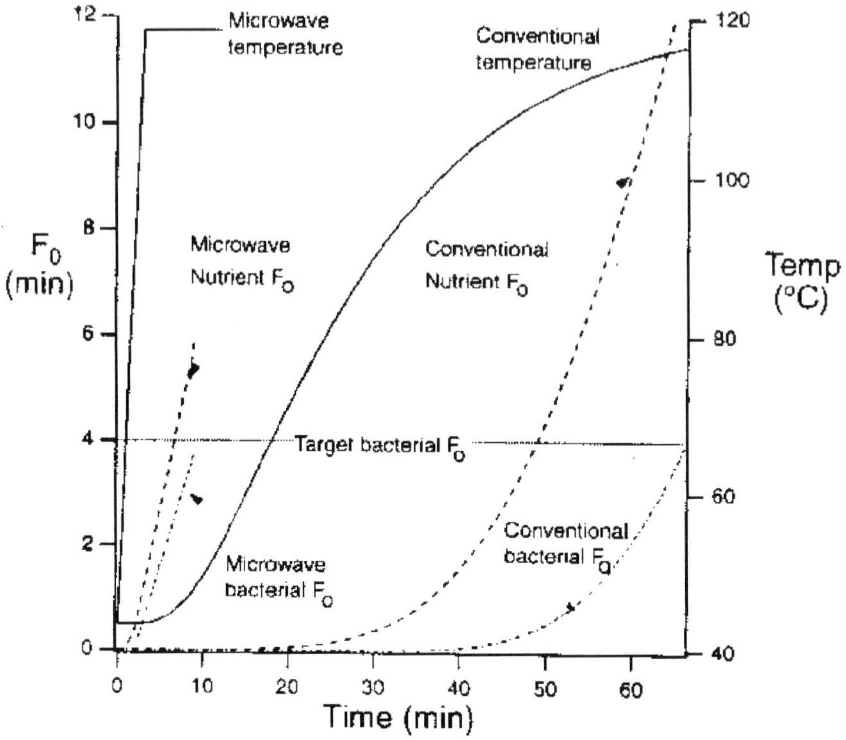

Figure 8.1 ***Quality parameters for microwave and conventional heating compared using computed values for typical heating situations, F_0 represents the accumulated lethality*** *(Source: Datta and Hu 1992).*

Microwave and radio frequency heating may be relatively more uniform than conventional heating, depending on the particular heating situation (Datta and Hu 1992); however, heating uniformity is hard to predict. Figure 8.2 illustrates a scenario in which microwave heating is spatially more uniform than conventional heating and helps demonstrate the reasoning behind it. The information shown in Fig. 8.2 is computed from mathematical models of a conventional and a comparable microwave heating process for a solid for input parameters given in Datta and Hu et al. Figure 8.2a shows that the range of temperatures reached by the 2 processes are approximately similar (as read from the horizontal axes) at the heating times shown.

The vertical axis shows the cumulative volume fractions of the food associated with a temperature, that is, for any temperature, the value on the curve signifies the volume fraction of food that has temperatures at or below this value. Figure 8.2b shows that the range of F_0 values (signifying time-temperature histories) are quite different for the same conventional and microwave heated food as in Figure 8.2a, for which temperatures are approximately similar.

The conventional heat process shows a much larger spread of F_0, which primarily signifies its tremendous non-uniformity of temperatures and long processing times leading to significant over-processing of the surface regions of the food. Other advantages of microwave and radio frequency heating systems are that they can be turned on or off instantly, and the product can be pasteurized after being packaged. Microwave and radio frequency processing systems also can be more energy efficient.

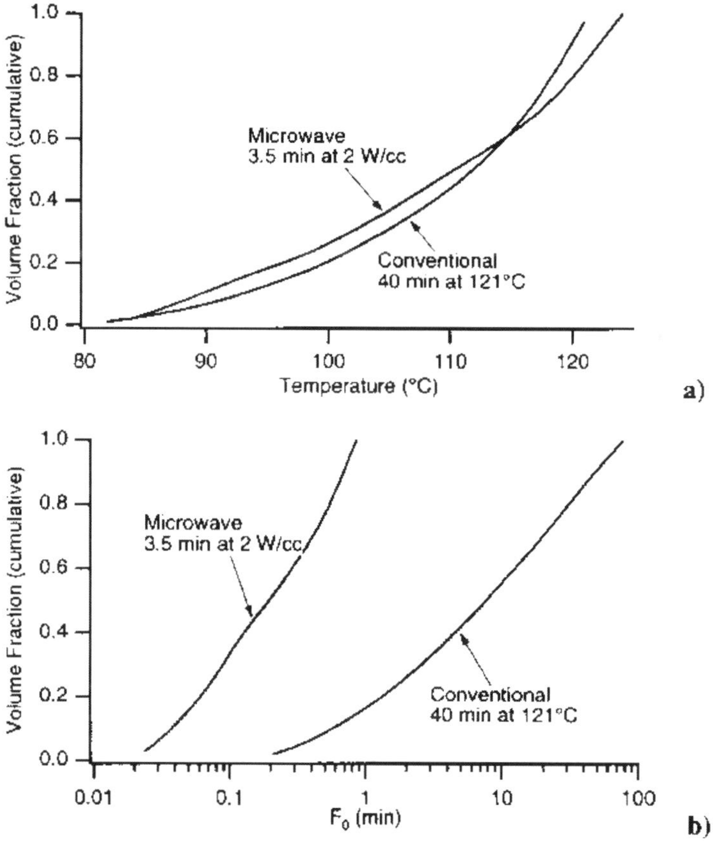

Figure 8.2 ***Illustration of how F_0 values (b) are typically quite different for microwave processing against conventional processing even when the range of temperatures are similar (a)*** *(Source: Datta and Hu 1992).*

Industrial Pasteurization and Sterilization Systems

Industrial microwave pasteurization and sterilization systems have been reported on and off for over 30 years (Tops 2000). Studies with implications for commercial pasteurization and sterilization have also appeared for many years. Early operational systems include batch processing of yogurt in cups and continuous processing of milk (Sale, 1976).

A very significant body of knowledge has been developed related to these processes. As of this writing, 2 commercial systems worldwide could be located that currently perform microwave pasteurization and/or sterilization of foods (Akiyama 2000; Tops 2000). As a specific example, ± company (Tops, 2000) produced over 13 million ready meals in 1998 and have installed a newly designed system in 1999. Although continuous microwave heating in a tube flow arrangement has been studied at the research level, no commercial system is known to exist for food processing.

Commercial radio frequency heating systems for the purpose of food pasteurization or sterilization are not known to be in use, although it has been researched over the y. The primary advantage of improved uniformity of heating was shown for in-package sterilization of foods in large packages using radio frequency at 27.12 MHz, although enhanced edge heating continued to be an issue.

Implementation of a microwave sterilization process can vary significantly among manufacturers. Unlike conventional heating, the design of the equipment can more dramatically influence the critical process parameter—the location and temperature of the coldest point. This uncertainty makes it more difficult to make general conclusions about processes, process deviations, and how to handle deviations. For example, in one implementation the process design consisted of heating, equilibration, holding, and cooling stages. The equilibration stage between heating and holding was to equilibrate the temperatures and avoid non-uniformities within the product. Hot air temperature and time are the factors controlling the equilibration process. All 4 stages are done under pressure to reach sterilization temperatures. The parameters recorded for the process were delivered power, temperature, pressure, speed, and cycle time.

Another system (Tops, 2000) consists of microwave tunnels with several launchers in relation to the number of products (ready meals). Microwave-transparent and heat-resistant trays are used with shapes adapted for microwave heating. Exact positioning of the package is made within the tunnel and the package receives a pre-calculated, spatially varying microwave power profile optimized for that package. The process consists of heating, holding and cooling in pressurized tunnels. The entire operation is highly automated (see monitoring of process deviations later).

For in-package pasteurization or sterilization, packaging materials need to be microwave transparent and have a high melting point. Also, because metal reflects microwaves, packages with some metal component can considerably change the food temperatures (critical process factor). In some situations, metals have been deliberately added to the package to redistribute microwave energy to achieve increased uniformity of heating. The most common packages that have been tried are individual pouches made of microwave transparent rigid films such as polypropylene with an ethylene vinyl alcohol (EVOH) barrier or a polyethylene terephtalate (CPET) film.

Shelf-life Extension at Home

Almost every US household owns a microwave oven that uses microwaves at a frequency of 2450 MHz. Reheating in a microwave oven, which the FDA does not regulate, is perhaps the most widespread use of the microwaves and has been known to involve serious microbiological safety issues. Researchers have also reported the use of home microwave ovens for pasteurization or for increasing shelf-life.

Future Processes

Many techniques have been tried to improve the uniformity of heating. These include rotating and oscillating the food package, providing an absorbing medium (such as hot water) surrounding the product, equilibrating after heating (Fakhouri and Ramaswamy 1993), and cycling the power. In the past, success of these processes has been limited due to the tremendous dependence of temperature and its distribution on food and oven factors. Use of the 915 MHz and radio frequencies to improve uniformity of heating may have potential for the future.

Future possibilities to improve the uniformity of heating include variable frequency microwave processing and phase control microwave processing. Although these 2 techniques have been applied to microwave heating of non-food materials, they are yet to be applied to food in any significant way. Combinations of microwave and conventional heating in many different configurations have also been used to improve heating uniformity. The critical process factor in combination heating or any other novel processes would most likely remain the temperature of the food at the cold point, primarily due to the complexity of the energy absorption and heat transfer processes. In the future, microwaves may be combined with conventional heating or chemical treatments for surface treatment, for example, meat processing or food contact surface.

Summary of Mechanism of Inactivation

The energy absorption from microwaves and radio frequency can raise the temperature of the food high enough to inactivate microorganisms for effective pasteurization or sterilization. A number of studies have proven that the thermal effect is the essential contributor to the destruction of microorganisms as well as the degradation of vitamin B_1, thiamin (Welt and Tong, 1993).

Since the beginning of microwave processing, there has been controversy about the possible non-thermal (also called "athermal") effects of microwave processing (these are effects unrelated to the lethality caused by the heat). Researchers have repeatedly reported non-thermal effects, although detailed discussions on those mechanisms are difficult to locate in the literature.

As many as 4 separate effects have been proposed: selective

heating of microorganisms, electroporation, cell membrane rupture, and cell lysis due to electromagnetic energy coupling. These mechanisms are discussed later in more detail; however, the general consensu is that the reported non-thermal effects are likely to be due to the lack of precise measurements of the time-temperature history and its spatial variations.

Summary of Microbial Inactivation Kinetics

Since the studies reporting non-thermal effects have been inconclusive, only thermal effects are presumed to exist. Thus, microbial inactivation kinetics for microwaves are essentially the same as the inactivation kinetics of conventional thermal processing. The microwave non-thermal effects have been reported to add to the destruction of microorganisms. Thus, ignoring the possible non-thermal effect can only provide an extra safety factor. To date, there do not appear to be any microwave-resistant foodborne pathogens.

Summary of Critical Process Factors

Since the thermal effect is the sole lethal mechanism assumed in this processing technology, time-temperature history at the coldest location will determine the safety of the process. Both the magnitude of time-temperature history and the location of the cold points are functions of the composition (ionic content, moisture, density, and specific heat), shape, and size of the food, the microwave frequency, and the applicator (oven) design. Time is also a factor in the sense that, as the food heats up, its microwave absorption properties can change significantly and the location of cold points can shift (See Figure 8.3).

Synergistic Effects

Microwave processes are sometimes combined with conventional heating. Synergistic effects, where the total effect of the combined process on the microorganism is more than the sum of individual effects of microwave and the other process, have not been reported.

Current Limitations / Status

As mentioned earlier, only 2 companies (Tops, 2000; Akiyama, 2000) could be located worldwide that are currently using microwave technology for pasteurization/sterilization of foods. Other systems may be operational, but details were not available.

Some reasons given for the lack of success in commercial operation are complexity, expense, non-uniformity of heating, inability to ensure sterilization of the entire package, lack of suitable packaging materials, and unfavorable economics when compared to prepared frozen foods in the United States. Current research at several universities (Washington State University, Cornell University) and a government agency (US Army Laboratories at Natick, MA) is aimed at further commercial use of microwave sterilization, particularly in the context of providing improved quality rations for soldiers.

Pathogens of Public Health Concern Most Resistant to the Technology

Identification of Pathogens Resistant to Microwaves

Numerous studies address the effect of microwave heating on pathogenic microorganisms in foods. Bacteria reported to be inactivated by microwave heating include *Bacillus cereus*, *Campylobacter jejuni, Clostridium perfringens,* pathogenic *Escherichia coli, Enterococcus, Listeria monocytogenes, Staphylococcus aureus,* and *Salmonella.* The nematode *Trichinella spiralis,* the organism that causes trichinosis, may also be inactivated. Foodborne pathogens have been shown to be inactivated by microwave heating in various poultry, beef, fish, and pork products, milk, and eggs; however, Heddleson and Doores, reported that "in-home pasteurization" of milk was "problematic" and "potentially dangerous" due to non-uniform heating and lack of standardization of home microwave ovens.

It is very difficult to precisely compare the effectiveness of microwave heating to conventional heating based on the literature, because of the different techniques employed or the lack of detail in the methods or materials used, especially in relation to temperature monitoring. A recurring conclusion in the literature is that non-uniform heating by microwaves may lead to survival of foodborne pathogens, including *Salmonella* and *L. monocytogenes,* in certain locations of foods heated at selected internal locations to endpoint temperatures that would normally be lethal.

For example, several studies have demonstrated that the measured internal temperature of poultry does not indicate the extent of inactivation of surface-inoculated *Salmonella* on poultry due to lower temperatures at the product surface. There do not appear to be any obvious "microwave-resistant" foodborne pathogens. Various studies have shown increased resistance of *S. aureus, C. perfringens,* or *Enterococcus faecalis* but not necessarily to the point that these could be labeled as resistant. As with conventional heating, bacteria are more resistant to thermal inactivation by microwave heating than yeasts and molds, and bacterial spores are more resistant than vegetative cells.

Effects of Critical Process Factors on Inactivation

As with other thermal processes, the primary factors that determine safety are temperature and time (that is, integrated time-temperature history). A number of critical process factors affect time-temperature history. These are discussed in detail in Chapter 4. Some of these critical process factors are moisture, ionic content, microwave frequency, product parameters (including mass, density, and geometry), specific heat, and the temperature achieved.

It is important to note that in the context of microwave processing, these critical process factors do not change the rate of inactivation per se. Rather; these factors change the spatial distribution of microwave absorption and, therefore, the spatially distinct heating rate and time-temperature history. The spatial distribution of time-temperature history, in turn, changes the distribution of the extent of inactivation within the food, thus generally changing the total inactivated population within a given food sample.

For example, the population of cells heated for 47 s at 700 W in a microwave oven in phosphate buffer were reduced by 99.8%, while those in 1% sodium chloride were reduced only by 62.4%. Such a difference is attributed to the effect of salt in decreasing the penetration of microwaves. Less microwave penetration leads to a lower internal temperature and a lesser destruction in the interior regions, resulting in an overall lower destruction.

Shape of Inactivation Curves

The shapes of the inactivation curves are expected to be similar to those for conventional heating.

Mechanisms of Inactivation

Pathogen Culture Maintenance

As stated above, microwave or radio frequency processing causes microbial inactivation predominantly through thermal effects. In reviewing the literature, no pathogen is identified as uniquely resistant to these processing methods. Therefore, maintenance of cultures (pathogen, surrogate vegetative cells, or spore crops) for evaluating the process or processing unit effectiveness should follow generally accepted culturing procedures for thermal process evaluation.

Conditions used for preparing, culturing, or storing vegetative cells or spores should be such that they produce the most resistant cell or spore. Appropriate conditions may be determined by consulting thermal resistance literature (See Chapter 4). Generally, specific conditions for the growth of the particular test microorganism should be defined. Cells incubated to stationary phase usually demonstrate maximum resistance. As suggested in the Overarching Principles Section, sublethal stress conditions also need to be evaluated, as they may increase resistance. As a rule, one should ensure that the test microorganism has a heat resistance equivalent to that generally recognized for the particular genus, species, and strain used.

Microbial Enumeration Conditions and Methods

Once the vegetative cell or spore is treated with microwaves, it must be enumerated to determine if it is still viable. The objective of the recovery process is to provide optimum conditions for treated cells or spores to grow to obtain a measure of the maximum number of non-injured and injured survivors (Overarching Principles Section).

For thermal processes, the length of incubation may be important in recovering viable cells or spores, because thermally treated cells or spores generally grow slower than non-treated ones. As with other process studies, experimentation will be necessary to determine the optimum conditions and methods for microbial enumeration.

Detailed Analysis of Inactivation Mechanisms

Two mechanisms are proposed for inactivation of microorganisms by microwaves. The first states that microwaves inactivate microorganisms entirely by heat through mechanisms comparable to other biophysical processes induced by heat, such as denaturation of enzymes, proteins, nucleic acids, or other vital components, as well as disruption of membranes. There is no question as to the validity of this mechanism.

A second proposed mechanism for inactivation by microwaves

involves non-thermal effects. Four predominant theories have been used to explain non-thermal inactivation by microwaves or "cold pasteurization": selective heating, electroporation, cell membrane rupture, and magnetic field coupling.

The selective heating theory states that solid microorganisms are heated more effectively by microwaves than the surrounding medium and are thus killed more readily.

Electroporation is caused when pores form in the membrane of the microorganisms due to electrical potential across the membrane, resulting in leakage.

Cell membrane rupture is related in that the voltage drop across the membrane causes it to rupture.

In the fourth theory, cell lysis occurs due to coupling of electromagnetic energy with critical molecules within the cells, disrupting internal components of the cell. These mechanisms have been studied extensively since the 1970's by a number of researchers.

Culkin and Fung reported earlier studies that suggested microwave heating at 2450 MHz caused greater destruction of *Aspergillus, Penicillium, Rhizopus*, aerobic microorganisms, *Salmonella* and *Proteus* in foods than heating alone. Culkin and Fung exposed *E. coli* and *Salmonella Typhimurium* in soups to 915 MHz microwaves and then determined survivors in the top, middle, and bottom regions of the product. The temperatures were measured using temperature-sensitive strips. They found that the greatest survival in the soups was in the top layer, which was also shown to have the lowest temperature.

A series of studies by Khalil and Villota suggested non-thermal effects of microwave heating. They first determined that *Bacillus stearothermophilus* spores in various media (water, NaCl, sucrose, phthalate, or phosphate buffers) had lower D100 °C values when 2450 MHz microwaves were used compared to using a heated water bath. The experiment appears to have involved 6 tubes at a single temperature with no replication. In addition, the come-up times, although a small part of the overall heating times (microwaves: 58–83 s out of 90–190 min, conventional heat: 100–135 s out of 113–240 min), were not considered.

Heddleson and Doores (1994) disputed the above conclusions due to inaccuracies in temperature measurement. Khalil and Villota (1988) further studied the effect of microwaves (2450 MHz assumed) on injury of *S. aureus* FRI-100. They heated cells at a sublethal temperature of 50 °C and maintained microwave temperature using recirculated cooled kerosene. Microwave heating caused a greater amount of cellular injury as determined by plating on trypticase soy agar plus 7% sodium chloride, increased loss of ultraviolet-absorbing cellular material, and extended time for enterotoxin production. Their findings also showed that microwave-injured cells recovered better when microwave heating was carried out anaerobically. This effect was not seen with conventional heating.

They speculated that the microwaves catalyzed oxidative reactions, possibly in membrane lipids, decreasing recovery of exposed cells. In another study, Khalil and Villota demonstrated that while both conventional and microwave heating destroyed the 16S subunit of RNA of sublethally-heated *S. aureus* FRI-100, only microwave heating affected the integral structure of the 23S subunit. Moreover, when cells were allowed to recover following injury, it took longer for the microwave treated cells to restore their 23S RNA.

Heddleson and Doores again concluded that these studies suffered from the lack of proper method of temperature measurement due to the unavailability of fiber optic thermometry. Kozempel and others designed a system in which various fluids were exposed to microwave energy (5.0–5.4 kW) and then cooled so as to maintain temperatures of the fluids at 40 °C. The fluids were inoculated with a bacterium reported to be *Pediococcus* strain NRRL B-2354 prior to exposure. The greatest kill took place in apple juice (up to 4.6-logs in ± pass), with moderate lethality (up to 0.7-logs in 1 pass), occurring in water and 10% glucose. The amount of kill with multiple passes through the system was not constant. In some products the kill rate was reduced following the first pass.

Little or no lethality was demonstrated with skim milk, pineapple juice, tomato juice, apple cider, or beer. In addition to the influence of the product itself, the medium used to grow the *Pediococcus* strain also appeared to affect cells counts; however, none of the product characteristics such as insoluble solids, pH, and conductivity could fully explain the variation. The authors concluded that they had demonstrated "significant microorganism kills in some fluids using microwave energy at sublethal temperatures."

Kozempel and others (2000) subsequently designed a new system that was capable of isolating thermal and non-thermal effects of microwave energy. The system was a double tube that allowed input of microwave energy but removed thermal energy with cooling water. With this system, the researchers found no inactivation of *Enterobacter aerogenes, E. coli, Listeria innocua, Pediococcus*, or a yeast in various fluids including water, egg white, whole egg, tomato juice or beer at sublethal temperatures. They concluded that, in the absence of other stresses such as pH or heat, microwave energy did not inactivate microorganisms; however, they did suggest that microwave energy may complement or magnify thermal effects.

In tests with *Saccharomyces cerevisiae* and *Lactobacillus plantarum* in apple juice, Ramaswamy and others (2000) also found that the non-thermal effect of microwave energy at sublethal temperatures is insignificant. However, they determined that, at equivalent heat treatments, microwaves enhanced inactivation. They demonstrated in a continuous flow system that *E. coli* K-12 had significantly lower D-values (12.98 s at 55 °C, 6.31 s at 60 °C, 0.78 s at 65 °C) using microwave energy than equivalent heat treatments with hot water (44.7 s at 55 °C, 26.8 s at 60 °C, 2.00 s at 65 °C) or steam (72.71 s at 55 °C, 15.61 s at 60 °C, 2.98 s at 65 °C). They concluded that, while there was no non-thermal effect of microwaves, there was a significant enhancement of thermal treatments.

Apart from the described studies, most research has concluded that there is little or no non-thermal effect of microwaves on microorganisms and that inactivation of microorganisms is due only to heat. Goldblith and Wang heated suspensions of *E. coli* and *Bacillus subtilis* under conventional heating and with microwaves at 2450 MHz. The degree of inactivation of both microorganisms was identical with conventional and microwave heating. Vela and Wu heated various bacteria, fungi, and bacteriophages in 2450 MHz microwaves in water and as lyophilized cultures. There was no inactivation of dry cultures even after extended exposure.

Similarly, Jeng and others found no difference in inactivation of *B. subtilis* spores under conventional or microwave (2450 MHz) heating in automated computer-controlled temperature monitoring systems. Kazbekov and Vyacheslavov found that thymidine and thymine uptake, leakage of potassium and hydrogen ions, and uptake of DNA by cells of *E. coli* or *B. subtilis* under low power microwaves were typical of that shown for heating.

Fujikawa and others found no major differences in inactivation kinetics of *E. coli* in phosphate buffer between microwaves and conventional heating. Welt and others demonstrated no difference between conventional and microwave inactivation of *Clostridium sporogenes* PA3679 at 90, 100, and 110 °C. A suspension of spores that was exposed to microwaves, but continuously cooled in silicone tubing demonstrated no detectable inactivation.

While there is some controversy as to the additional inactivation of microorganisms over the thermal effect of microwaves, this additional inactivation is small and inconsistent. In many studies comparing microwave heating to conventional heating, microwave heating appears less effective due to nonuniform heating effects from unpredictability of cold spots and changing product parameters, such as specific heat. Therefore, when developing methods for describing the inactivation kinetics of microwave heating, it is recommended that only thermal effects be included in the model.

Under the assumption of only thermal effects, the kinetic parameters presented in Table 10-1a of the Overarching Principles are recommended for use in design of processes involving the microwave treatment of foods. The kinetic parameters used to design thermal processes have been presented and defined in an introductory chapter of the main document. The pathogens of concern will be the ones defined for thermal processing and described in the Overarching Principles of this report.

Validation/Critical Process Factors

Identification and Description of Critical Process Factors

Time-temperature history at the coldest point determines the microbiological safety of the process, as in other thermal processing. Once temperature is known at the coldest point as a function of time, accumulated lethality can be calculated following the well-known equation 8.2:

$$F_0 = \int_0^{V} 10^{(\Gamma - 250)/z} dt \qquad \text{Eq. 8.2}$$

where T is the cold point temperature at any time t, z is the z-value in F_0 and t_f is the total duration of heating. There are, however, major differences between conventional and microwave heating in terms of the location of the cold point and how time-temperature history of the cold point is affected by a number of critical process factors. ***Note that the effect of the factors are discussed in a simplistic way in order to illustrate the concepts—the actual influence of the factors can be quite complex and are only known from detailed experiment or mathematical modeling.*** Such effects are discussed in detail in books such as Datta and Anantheswaran (2000), Buffler and Decareau or review papers such as Saltiel and Datta.

Time-temperature history at the coldest point for a conventional thermal process is generally quite predictable for a food that is all solid or all fluid. For example, for a conduction-heated (solid) food, it is usually the geometric center. In microwave heating, even for a solid food, the coldest point is less straightforward to predict and can change during the heating process (Figure 8.3), depending on a number of food and oven factors.

Accordingly, relatively sophisticated modeling based on measured properties of the foods needs to be used and subsequently validated to ascertain the location of the point of lowest integrated time-temperature history. Well-developed but simple procedures, such as the Ball formula would be much harder to achieve for microwave heating. Changes in properties during heating have a more pronounced effect in microwave heating as compared to conventional heating. As the food heats, its microwave absorption capability typically increases, which increases the rate of temperature rise and therefore further increases the rate of microwave absorption.

Such coupling could lead to runaway heating. Figure 8.3 also illustrates the coupling effect. Initially, at lower temperatures, microwave absorption is lower, so the waves are able to penetrate a lot further into the material. As the material heats up, it absorbs microwaves more readily and the waves are not able to penetrate as far. Especially in foods with high ionic concentrations, the surface at higher temperatures can act as a shield.

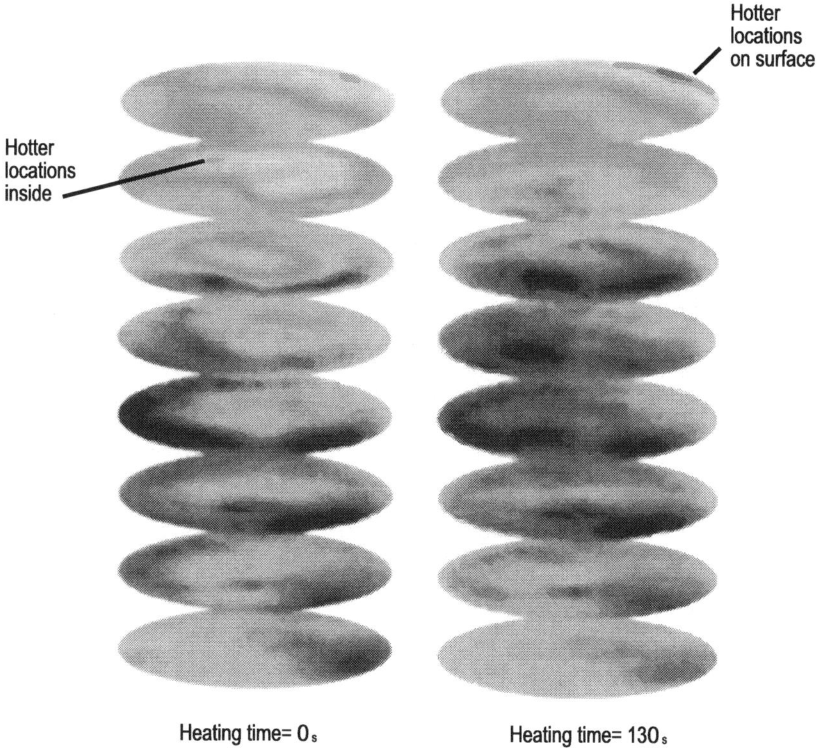

Figure 8.3 ***Microwave power absorption (Q in equation 8.1) patterns in a sterilization process can change dramatically during heating, as shown by the migration of hottest locations (in red) from interior (a) to surface (b). Shown are computed results for a ham cylinder (0.7% salt) heated in a microwave oven similar to domestic microwave ovens*** *(Source: Zhang and others 1999).*

Since heat is constantly generated everywhere in the food, but at different rates, the difference between the temperature at the coldest and the warmest points in the food keeps increasing with time. This is unlike conventional heating where the coldest point approaches the warmest temperature of the system (typically the heating medium temperature) with time. In conventional heating, the surface is at the highest temperature, corresponding to the temperature of the heating medium. In microwave heating, the food heats up while the surrounding air stays cold (Datta 2000).

The cold air keeps the surface temperature lower than locations near the surface of food. Surface evaporation, especially when heating an unpackaged food, can further decrease the surface temperature. In some heating applications, such as with frozen foods that are spherical, the surface could be the coldest location. In conventional heating, the maximum temperature is limited by the heating medium temperature, such as steam in a retort. Since microwave absortion continuously generates heat, temperature keeps increasing in the microwave heating process. To keep the temperature within reasonable limits, microwaves need to be turned on and off (cycled) once the target temperature has been reached.

One of the advantages of microwave heating is that the come-up time is short. It is this shorter come-up time that helps retain the organoleptic qualities and that is the basis for preferring microwave processing to conventional thermoprocessing. In calculating the process time, the come-up time in microwave heating cannot be given nearly as much importance as in conventional heating (See Fig. 8.1).

Factors Related to Product and Package

Food shape, volume, surface area, and composition are critical factors in microwave heating. These factors can affect the amount and spatial pattern of absorbed energy, leading to effects such as corner and edge overheating, focusing, and resonance. For example, a curved shape can focus microwaves and produce a higher internal rate of heating than near the surface. Such heating patterns can also change with time, as illustrated in Figure. 8.3. The effect of food volume on total amount of energy absorbed by the food for a given setting of power level is typically as shown in Figure 8.4. Since the total energy absorbed lags the increase in volume, average temperature rise drops (however, food as a whole heats slower).

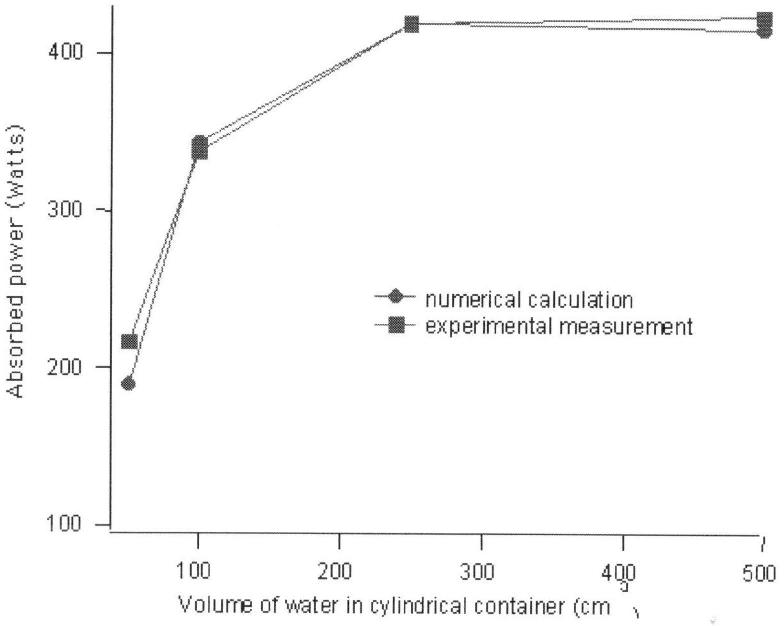

Figure 8.4 ***Magnitude of Total Absorbed Power as a Function of Volume of Food, Obtained from Experiments and Electromagnetic Simulations***
(Source: Zhang and others, 1999).

Composition, in particular moisture and salt percentages, has a much greater influence on microwave processing than in conventional processing, due to its influence on dielectric properties. High salt and moisture content increases the efficiency of microwave absorption, thereby decreasing the depth of penetration. Thus, interior locations generally get heated less in foods with high salt or moisture content, reducing microbial destruction.

Composition can also change the thermal properties such as specific heat, density, and thermal conductivity and, thereby, change the magnitude and uniformity of the temperature rise. For example, the temperature of a low specific heat oil increases at a much faster rate than that of water when compared at the same level of absorbed power. The different components of mixed food products, such as multi-compartment frozen dinners, will heat differently. Packaging material is also a critical process factor. In contrast to commercial canning, where metal containers offer minimum thermal resistance and are not a critical process factor, metallic components present in a package, such as aluminum foil and susceptors, can dramatically influence the heating rates of the packaged food.

Factors Related to Process and Equipment

Several process and equipment factors are critical in microwave heating. Design (size, geometry, and so on) of the microwave oven can significantly affect the magnitude and/or spatial variation of the power absorption in the product. In addition, presence or absence of devices added to improve uniformity, such as mode stirrers and turntables, are major factors affecting temperature distribution.

The placement of the food inside the oven can also have a significant influence on the magnitude and uniformity of power absorption. Other factors related to the equipment are the temperature of the medium surrounding the product and the level of food surface evaporation (especially significant for unpackaged food), both affecting food surface temperature.

Cooling effects due to surface evaporation have been shown to lead to survival of *Trichina larvae* in pork. Heating of containerized liquids with microwaves without agitation causes flow and thermal stratification inside the container. Warmer liquid moves to the top, much like in conventional heating. Due to variations in product characteristics (such as viscosity), in package components (such as metal in aluminium foil), and in equipment factors, the patterns of temperature distribution in heated liquids (static or flowing) can be quite complex, and the slowest heating location needs to be determined for distinctly different situations.

In addition to the aspects discussed above, power level and cycling of the microwave input are critical process factors in microwave heating. Also, power output by the magnetron (the component in an oven generating the microwaves) changes as the magnetron heats up over time. Thus, equipment specific "wait time" may be necessary before the power output becomes stable. Due to differences in penetrating ability, the frequency of the microwaves can dramatically affect the heating rates and their spatial distribution. In a simplified view, a lower frequency of 915 MHz has a higher depth of penetration than the 2450 MHz used for home microwaves. At this lower frequency, uniformity of heating can improve with reduced edge heating (Lau and others, 1998).

Equilibration of the product following heating can help to level the temperature distribution and improve uniformity. Its important effects have been demonstrated. A summary of the various product, package, process, and equipment factors discussed above is provided in Table 8-2. Due to the number of critical factors implicated, none of them alone can be treated as a critical process factor by itself, unless all others are held constant.

Table 8-2 *Critical Process Factors in Microwave Heating*

Food	Shape, size, composition (moisture, salt, and so on), multiple components (as in a frozen dinner), liquid against solid.
Package	Presence of metallic elements such as aluminum foil, susceptor.
Process	Power level, cycling, presence of hot water or air around the food, equilibration time.
Equipment	Dimensions, shape and other electromagnetic characteristics of the oven, frequency, agitation of the food, presence of mode stirrers and turntables.

Identification of the Effect of Process Factors on Cold Point Using Mathematical Simulation

Due to the complexity of the system where the heating pattern depends on such a large number of factors, simulation-based design can save significant time and resources in developing microbiologically safe processes. Such simulation-based design can drastically reduce the number of experiments needed to predict the location of cold points and the time-temperature history at these locations for the actual food and equipment combinations. State of the art commercial software simulating the electromagnetic and heat transfer properties has been used for microwave food process design (Dibben 2000; Zhang and Datta 2000). Such software can provide a comprehensive insight into the heating process by showing interior power absorption (that is, heating rates) in a 3D object (Figure 8.3), difficult using experimentation.

Simulation-based design can allow the process and equipment designers to judiciously choose proper combinations of food and process parameters in an efficient manner, reducing some of the time and expenses in prototype building. Location of coldest point (the critical process factor) and its time-temperature history can be predicted this way. As the user friendliness, accuracy, and linkages with other software improve, more food processors are expected to use these programs routinely.

Description of Methods to Measure / Monitor Critical Process Factors

Monitoring the temperature of microwave processed food poses a challenge. Thermocouples and other metallic probes used to record temperatures for conventional processing in static systems are generally unsatisfactory for precision temperature measurements in microwave ovens for several reasons. Firstly, metallic probes reflect and absorb the energy of incident microwaves, and require special grounding and installation to withstand microwave operations. Secondly, electromagnetic field disturbances caused by the presence of metallic probes create localized changes in heating patterns that can produce variability in overall heating patterns.

Valuable alternatives to metallic probes, however, are fiber-optic temperature probes. These are on the market and have been used to monitor temperatures during microwave heating. They are inert to the electric and magnetic fields of the microwaves or radio frequencies. Additional advantages are their accuracy (from 0.2 to 0.05 ^{0}C) and fast response (milliseconds). Some disadvantages of fiber-optic probes are their current price (although it is dropping) and somewhat fragile nature.

Description of Microbial or Chemical Surrogates / Indicators

For determining the kinetics and efficiency of microwave inactivation of microorganisms, surrogate/indicator microorganisms could be selected from those traditionally used in thermal processing studies. No microorganisms with unique resistance to microwave processing have been reported in the literature, suggesting that classical surrogates (vegetative cells or spores) would be appropriate for process determination and validation.

Since microwave or radio frequency processing is primarily a heat process, microbiological validation tests should be designed using procedures that parallel those used for thermal processing. Important considerations for an inoculated pack study of a new thermal process include selection of a surrogate microorganism, preparation and handling of the test microorganism, size and volume of inoculum, method of inoculation, processing levels and conditions, number of containers, product data collection, statistical techniques and methods for determining survivors (for example, incubation and microbiological recovery).

A more detailed discussion of these procedural considerations is available in the Laboratory Manual for Food Canners and Processors. The principles and methods are similar, whether the objective is pasteurization, pathogen reduction, or commercial sterilization. One problem that needs to be carefully considered is the method of inoculation. Since microwave heating can be nonuniform and the cold spot is not easy to identify, the inoculum should be distributed throughout the food product. This could be a particular problem with solid products, unless made homogeneous by grinding or placed in locations where significant spatial variation in heating rates is expected.

The use of history indicators or time-temperature integrators, either biological (microbiological or enzymatic) or chemical, could be a way to monitor the process impact and could be helpful in identifying critical process parameters. As in thermal processing, proper calibration of the kinetic parameters of the surrogates/indicators is required. The same principles as for thermal processing apply and extra care should be given to critical factors associated with the microwave/radio frequency heating process in that the presence of the indicator should not influence the heating process.

An example is the validation of sterilization patterns by correlating thermally induced chemical changes in the food (a history indicator) to bacterial destruction. Intrinsic chemical markers whose extent of formation is a function of time-temperature history have been used recently in several processing situations, including microwave sterilization. This approach can provide information on spatial distribution of the integrated time-temperature history within a packaged system and on any variation among packages in a continuous process operation.

Process Deviations

Process deviations in microwave processing present some special issues and challenges. Temperatures are generally more difficult to monitor and measurement of power output from microwave generators may not accurately reflect product temperature, unless the sensitivity of the heating process to changes in food composition, size, shape, placement in the oven and other factors discussed earlier are taken into consideration.

Due to the complex nature of the process, adjustments such as extending the time or increasing the power level will not be simple. It is generally believed that complete reprocessing would be the most reliable way to handle underprocessed material.

Basic Detection Methods for Process Deviations

Process parameters under direct control of the operator are the power level (including cycling), spatial distribution of power (number and positioning when multiple microwave generators are used) and time of exposure. As mentioned above, once a deviation has taken place, adjustments could be complex. In one implementation (Tops, 2000), the exact placement of a product in the tunnel is known and power levels of multiple microwave generators are programmed precisely to provide the custom-tailored heating profile for that tray and product. For example, the center of the tray is provided with higher microwave power.

The detection system consists of a variety of monitors. For example, broken generators and insufficient power level delivered by a generator is automatically recorded. Infrared surface measurement of each tray can be made while they are being transported to the hold system. Swelling of the top surface of individual packages due to internal steam generation during heating is monitored using a distance tracer—adequate heating produces enough steam for the package to swell sufficiently. Visual control is also made by placing maxi-thermometers (that measure maximum temperatures) at precise locations in the package.

An endoscope is also used to observe the heating process inside the microwave tunnel and therefore to monitor it manually. In another implementation, power settings for individual magnetrons are stored over time. If the power delivered varies from the set values, an alarm warns the operator. An additional warning signal comes from any blockage of the product feeding system that may lead to unintended cooling of the preheated products. This publication also reported the use of a Data Trace metallic temperature data-logger inside the package by ± company to monitor the time-temperature history. Use of such metallic data-loggers requires careful considerations and interpretations.

Methods to Assess and Correct Deviations

Details about process deviations are hard to obtain. Only ± company was willing to share information about current production of microwave pasteurized and sterilized food (Tops, 2000). In this company's implementation, a product is rejected based on an automated control system. Rejection is done at the end of the cooling system based on infrared surface temperature measurement, detection of broken microwave generators, and other means. Further loading of food into that tunnel is automatically stopped following a rejection. The temperature control system for each microwave tunnel can also be adjusted, if necessary, following a deviation. The control system is also programmed for each individual product. In general, extensive experimentation would be needed to validate the effectiveness and reliability of the methods to assess and correct deviations.

References

Anantheswaran, R. C. and Liu, L. Z. "Effect of viscosity and salt concentration on microwave heating of model non-Newtonian liquid foods in cylindrical containers." *J Microwave Power Electromagnetic Energy*. 29a (1994) : 127. Print.

Anantheswaran, R. C. and Liu, L. Z. "Effect of shielding on flow profiles in water in a cylindrical container during microwave heating." *J Microwave Power Electromagnetic Energy*. 29ab(4) (1994) : 220. Print.

Anonymous. "The potential of Bach". *Food Manuf.* 55(10) (1980) : 53. Print.

Anonymous. "Sterilization surfaces by irradiation with microwaves." *NASA Tech Briefs.* (1996.) : 140. Print.

Anonymous. Micro Denshi Co., Ltd., No. 8-25, 2-Chrome Yoshinodai, Kawagoeshi, Saitama 350, Japan. (1999). Print.

Ball, C. O.; and Olson, F. C. W. *Sterilization in Food Technology.* McGraw-Hill Book Company. New York. (1957). Print.

Bassani, M. Personal communication. Micromac Systems. (1999). Print.

Bengtsson, N. E.; and Green, W. "Radio-frequency pasteurization of cured hams." *J Food Sci.* 35 (1970) : 681-687. Print.

Buffler, C. R. *Microwave cooking and processing: Engineering fundamentals for the food scientist.* Van Nostrand Reinhold. New York. (1993). Print.

Burfoot, D.; Griffin, W. J.; and James, S. J. "Microwave pasteurization of prepared meals." *Journal of Food Engineering.* 8 (1988) : 145-156. Print.

Burfoot, D.; Railton, C. J.; Foster, A. M.; and Reavell, R. Modeling the pasteurization of prepared meal with microwaves at 896 MHz. *J Food Eng.* 30 (1996) : 117-133. Print.

Burton, H. "A survey of literature on bacterial effects of short electromagnetic waves." Shinfield, England. National Institute for Research in Dairying Shinfield. N. I. R. D. Paper No. 1041. (1949). Print.

Casasnovas, J.; Anantheswaran, R. C.; Shenk, J.; and Puri, V. M. "Thermal processing of food packaging waste using microwave heating." *J Microwave Power Electromagnetic Energy.* 29 (1994) : 171. Print.

Chipley, J. R., Effects of microwave irradiation on microorganisms. Adv Appl Microbiol. Academic Press Inc. 26. (1980) : 129-145. Print.

Chiu, C. P.; Tateishi, K.; Kosikowski, F. V.; and Armbruster, G. Microwave treatment of pasteurized milk. J Microwave Power. 19(4) (1984) : 269-272. Print.

Cross, G. A.; and Fung, D. Y. C. "The effect of microwaves on nutrient value of foods." *CRC Crit Rev Food Sci Nutr.* 16 (1982) : 355-381. Print.

Culkin, K. A.; and Fung, D. Y. C. "Destruction of *Escherichia coli* and *Salmonella typhimurium* in microwave-cooked soups." *J. Milk Food Technol.* 38(1) 1975) : 8-15. Print.

Datta, A. K. "Mathematical modeling of microwave processing as a tool to study safety". *American Society for Agricultural Engineers Paper,* 91 (1991) : 6614. Print.

Datta, A. K.; and Hu, W. "Quality optimization of dielectric heating processes." *Food Technol.* 46(12) (1992) : 53-56. Print.

Datta, A. K.; and Liu, J. "Thermal time distributions for microwave and conventional heating of food." *Trans I Chem E.* 70(C) (1992) : 83-90. Print.

Datta, A. K.; Sun, E.; and Solis, A. "Food dielectric property data and its composition-based prediction." M. A. Rao and S. S. H. Rizvi (eds.). *Engineering Properties of Food. New York.* Marcel Dekker. (1994.) : 457-494. Print.

Datta, A. K. Fundamentals of heat and moisture transport for microwaveable food product and process development. A. K. Datta and R. C. Anatheswaran. (eds.) (2000). Print.

Handbook of Microwave Technology for Food Applications. Marcel Dekker, Inc. New York. (2000). Print.

Decareau, R. V. *Pasteurization and sterilization. Microwaves in the food processing industry.* Academic Press. (1985). Print.

Dibben, D. "Electromagnetics: fundamental aspects and numerical modeling." A. K. Datta and Anatheswaran (eds.). *Handbook of Microwave Technology For Food.* Marcel Dekker, Inc. New York. (2000). Print.

Fakhouri, M. O. and Ramaswamy, H. S. "Temperature uniformity of microwave heated foods as influenced by product type and composition." *Food Res Int.* 26 (1993) : 89-95. Print.

Fleischman, G. J. "Predicting temperature range in food slabs undergoing long term low power microwave heating." *J Food Eng.* 27(4) (1996) : 337-351. Print.

Fujikawa, H.; Ushioda, H.; and Kudo, Y. "Kinetics of Escherichia coli destruction by microwave irradiation." *Appl Environ Microbiol.* 58(3) (1992) : 920-924. Print.

Fung, D. Y. C. and Cunningham, F. E. "Effect of microwaves on microorganisms in foods." *J Food Protect.* 43(8) (1980.) : 641-650. Print.

Goldblith, S. A. and Wang, D. I. C. "Effect of microwaves on Escherichia coli and Bacillus subtilis." *Appl Microbiol.* 15(6) (1967) : 1371-1375. Print.

Hamid, M. A. K.; Boulanger, R. J.; Tong, S. C.; Gallop, R. A.; and Pereira, R. R. "Microwave pasteurization of raw milk." *J Microwave Power.* 4(4) (1969) : 272-275. Print.

Harlfinger, L. "Microwave sterilization." Food Technol. 46(12) (1992.) : 57-61. Print.

Heddleson, R. A. and Doores, S. "Factors affecting microwave heating of foods and microwave induced destruction of foodborne pathogens" , (A review) *J Food Protect.* 57(11) (1994) :1025-1037. Print.

Chapter 9

Gas, Cold Plasma Technology

Introduction

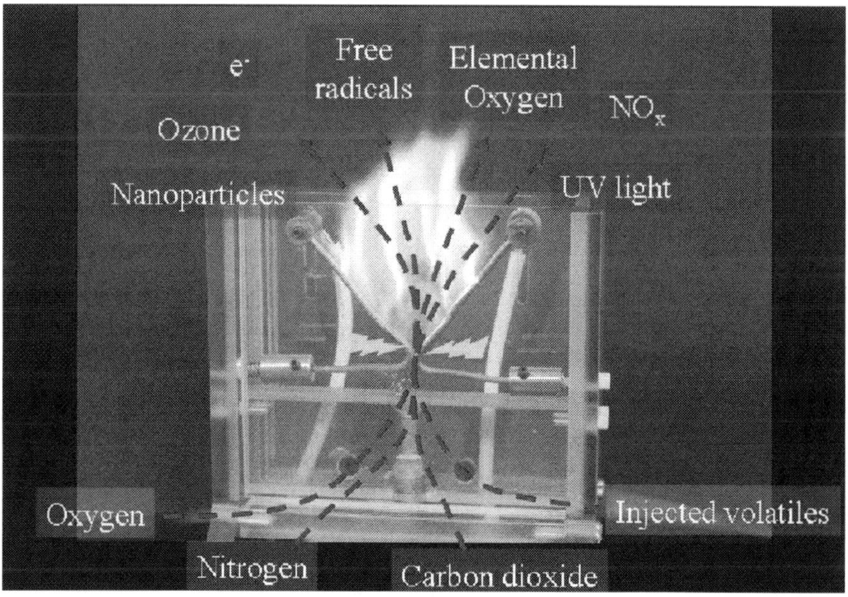

Figure 9.1 ***Illustration of Components and Creation of Cold Plasma***
(Source: Cheng, 2006)

Beyond UV-light and ozone treatments, a very novel and future processing method is cold plasma treatment. Very little information is currently available in the literature concerning cold plasma. Currently, it is a surface treatment technique used for modifying surface properties of polymers (Cheng et al., 2006; Groning et al., 2001). Typically, these modifications are aimed:

- To modify the surface energy *to improve adhesion between different polymers or between an aluminum coating and a polymeric substrate.*

- To modify the chemistry of surface *to improve polymer functionality, as for example the water vapor barrier property of poly(L-lactic acid) film (Uemura et al., 2006) or the inertia property of PVC films (Audic et al., 2001).*

A new challenge is related to the possibility of using cold plasma as a novel route to sterilize packaging materials (Schneider et al., 2005). However, the effects of cold plasma on packaging safety aspects as well as on the structural and functional properties of packaging materials should be understood first and then quantified. Although information is not yet available on this subject, a quick overview of literature papers dealing with cold plasma applications in the packaging science field can, however, give some information on the behavior of common plastic materials under cold plasma processing.

Cold plasma is a partly ionized gas, which is generated in a very intense electric field in a low pressure (typically $0.1 \leq p \leq 103$ Pa). Electrons, which acquire their kinetic energy from the applied electromagnetic field, collide with gas molecules and thereby transfer energy through ionization, bond breakage (molecular fragmentation), and others forms of excitation. Excited molecular and atomic species in the plasma, in turn, can emit photons over a very broad portion of the electromagnetic spectrum, ranging from X-rays to the infrared (IR).

Photons in the vacuum ultraviolet (VUV) or ultraviolet (UV) range, which have energies comparable or exceeding those of covalent bonds (a few eV) can readily initiate photo-chemical reactions in the polymers. Actions of plasma result in cross-linking, functional group formation, and degradation (chain scission). The relative importance of each phenomenon depends on the plasma generating conditions, the polymer composition, and the plasmagenic gas (Audic et al., 2001).

Different types of gases such as argon, oxygen, nitrogen, fluorine, water etc., can be used. For example, O_2 plasma treatment can increase the surface energy of polymers, whereas fluorine-containing plasma treatment can decrease the surface energy and improve the chemical inertness. Cross-linking of flexible films was expected to prevent the migration of additives and other low-molecular substances as monomers since the tortuosity of the diffusion pathway is increased. On the contrary, degradation induces etching and the formation of breakdown products and increasing migration of low-molecular substances should be then expected.

In order to select the most efficient gas to favor cross-linking and, thus, prevent migration of additives by plasma treatment, Audic et al. (2001) have carried out a large study using PVC film with di-2-ethylhexyladipate (DEHA) and epoxidized soybean oil (ESO) as PVC plasticizers and N_2, CO_2, He, H_2, and Ar as plasmagenic gas.

Argon plasma treatment with plasticized PVC films led to samples that exhibited a very low migration rate of DEHA and ESO: 5 min of treatment caused a migration rate decrease from 15.8% to the lowest observed value (3%). The following gases in the classification were He and H_2. The less efficient plasmagenic gases were CO_2 and especially N_2; 5 min of N_2 plasma treatment caused a migration rate decrease from 15.8% to only 11%.

The good results obtained with Ar and He could be explained by the fact that inert gas plasmas are known to induce polymer cross-linking. Ar plasma was shown to give higher radiative energy transfer than He plasma, consequently inducing more cross-linking than He plasma. H_2 plasma was shown to induce cross-linking in plasticized PVC. On the contrary, N_2 is known to preferentially induce the linking of functional groups and CO_2 is known to preferentially induce both degrading (chain scission) and fictionalisation (Brosse et al., 1997).

Besides the surface functional group formation or cross-linking induction, the plasma treatments are known to induce ablation of matter at the surface, which could be monitored by weight loss measurements. Regarding the weight loss of treated plasticized PVC foils in the study of Audic et al. (2001), the N_2 plasma treated films were less degraded than the Ar one, followed by He, H_2, and CO_2. Indeed, if Ar induces high cross-linking in PVC, this cross-linking is not directly correlated to a low degradation. It is a crucial point since this degradation leads to the breakdown of a small number of molecules into low molecular weight components, which could be eventually vaporized and pumped out of the plasma reactor but could also remain in the film and further migrate in the food upon contact.

If Ar plasma treatments are suitable for preventing migration of additives such as DEHA and ESO on one side by cross-linking, they may favour on the other side the apparition of new, non-characterized degradation products which could be worth being characterized in further studies. However, according to Audic et al. (2001), with a very low additives migration rate and a reasonable weight loss, the Ar plasma offers the best balance between cross-linking and degradation. In light of the few studies cited below, cold plasma treatment appears as a very promising technology for further research on packaging surface cleaning and disinfection.

Indeed, through the cross-linking and degradation phenomena induced in synthetical polymers one can expect a similar effect on bio-macromolecules. However, cold plasma technology is still in its infancy, and the plasma chemical process is not fully understood. Adjustment and optimization of the technique should be processed as a function of the targeted effects on polymers. Besides, the effects of cold plasma on additives migration out of packaging into food or breakdown products formation are strongly dependent on the processing conditions and should be investigated in detail.

Applications and Technologies

In the context of food processing, non-thermal plasma (NTP) is specifically an antimicrobial treatment being investigated for application to fruits, vegetables and other foods with fragile surfaces. These foods are either not adequately sanitized or are otherwise unsuitable for treatment with chemicals, heat or other conventional food processing tools. The term cold plasma has been recently used as a convenient descriptor to distinguish the one-atmosphere, near room temperature plasma discharges from other plasmas, operating at hundreds or thousands of degrees above ambient (see here for more on plasma temperatures). Within the context of food processing the term "cold" can potentially engender misleading images of refrigeration requirements as a part of the plasma treatment. However, in practice this confusion has not been an issue.

The nomenclature for non-thermal plasma found in the scientific literature is varied. In some cases, the plasma is referred to by the specific technology used to generate it ("gliding arc", "plasma pencil", "plasma needle", "plasma jet", "dielectric barrier discharge", etc.), while other names are more generally descriptive, based on the characteristics of the plasma generated ("one atmosphere uniform glow discharge plasma", "atmospheric plasma", "ambient pressure non-thermal discharges", "non-equilibrium atmospheric pressure plasmas", etc.). The two features which distinguish NTP from other mature, industrially applied plasma technologies, is that they are:

- *Non-thermal.*
- *Operate at or near atmospheric pressure.*

Table 9-1 *NTP Technologies*
(Source: Author's compilation)

	NTP Technology Class		
	I. Remote Treatment	**II. Direct Treatment**	**III. Electrode Contact**
Nature of NTP applied	**Decaying plasma** (afterglow) Longer lived chemical species.	**Active plasma** Short and long-lived species.	**Active plasma** All chemical species, including shortest lived and ion bombardment.
NTP density and energy	**Moderate density** Target remote from electrodes. However, a larger volume of NTP can be generated using multiple electrodes.	**Higher density** Target in the direct path of a flow of active NTP.	**Highest density** Target within NTP generation field.
Spacing of target from NTP-generating electrode	**Approx. 5 – 20 cm** Arcing (filamentous discharge) unlikely to contact target at any power setting.	**Approx. 1 – 5 cm** Arcing can occur at higher power settings, can contact target.	**Approx. ≤ 1 cm** Arcing can occur between electrodes and target at higher power settings
Electrical conduction through target	No	Not under normal operation, but possible during arcing.	Yes, if target is used as an electrode OR if target between mounted electrodes is electrically conductive.
Suitability for irregular surfaces	**High** Remote nature of NTP generation means maximum flexibility of application of NTP afterglow stream.	**Moderately high** NTP is conveyed to target in a directional manner, requiring either rotation of target or multiple NTP emitters.	**Moderately low** Close spacing is required to maintain NTP uniformity. However, electrodes can be shaped to fit a defined, consistent surface.
Examples of technologies	Remote exposure reactor, plasma pencil.	Gliding arc; plasma needle; microwave-induced plasma tube.	Parallel plate reactor; needle-plate reactor; resistive barrier discharge; dielectric barrier discharge.

Enclosed Plasma Treatment Chambers

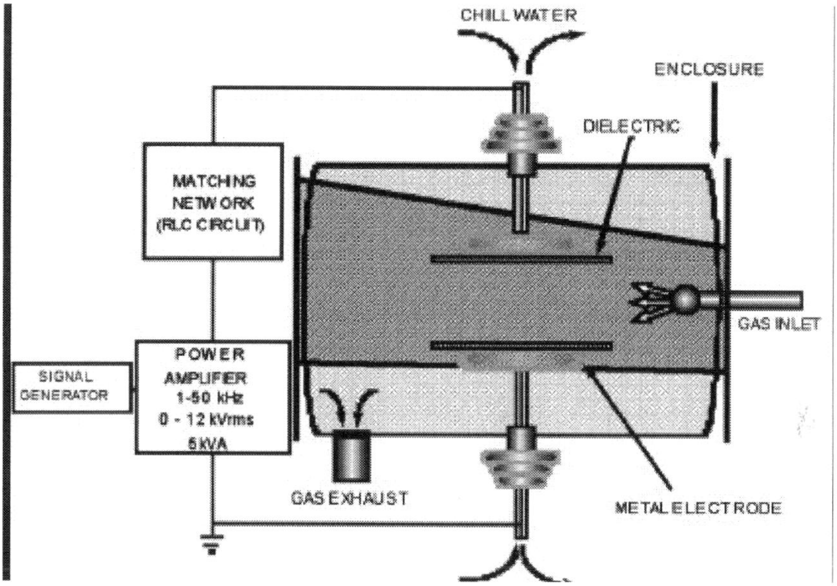

Figure 9.1 ***Enclosed Plasma Treatment Chambers***
(Source: OAUGDP. Kayes, M.M. et al., 2007. Foodborne Path Dis 4(1). DOI: 10.1089/fpd.2006.62)

Figure 9.2 ***Microwave Pumped Plasma Enclosed Chamber***
(Source: Amidi, M., et al. 2007. Food Science Australia)

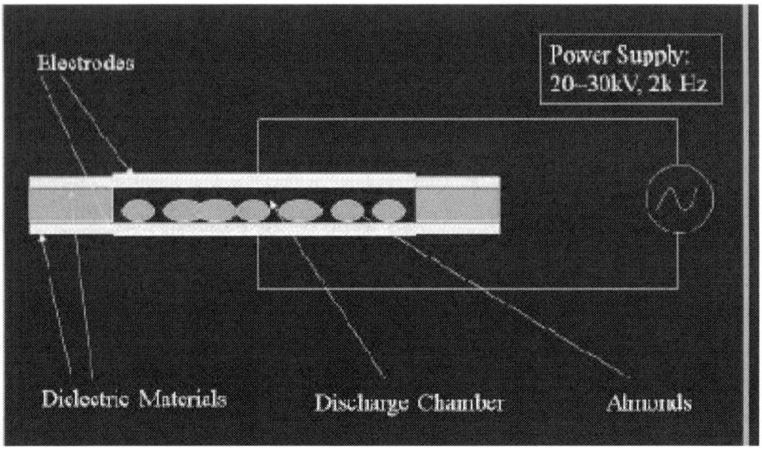

Figure 9.3 ***Electrode Contact Plasma Treatment***
(Source: Amidi, M., et al. 2007. Food Science Australia fpd.2006.62)

Direct Application of Plasma, Carrier Gas

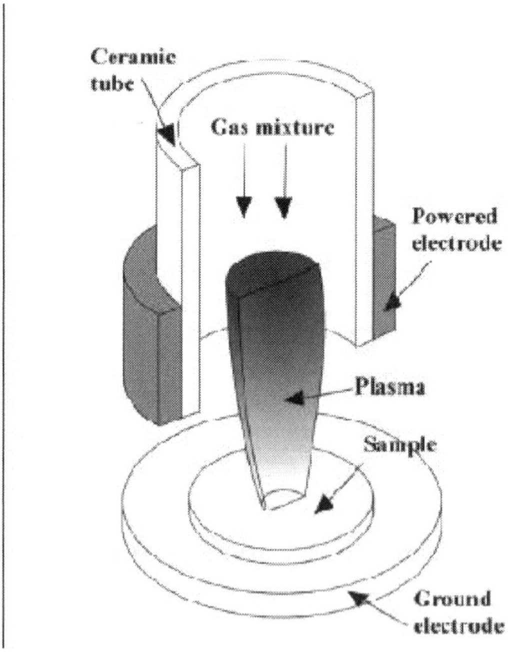

Figure 9.4 ***Schema of the Cold Atmospheric Gas Plasma Rig***
(Source: Perni, S. et al. JFP, 71(2) (2008) : 302–308)

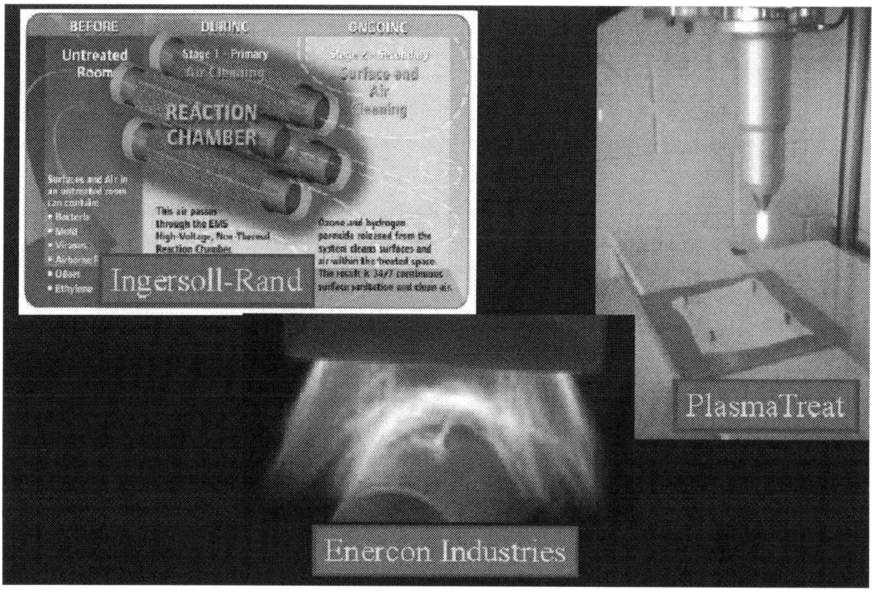

Figure 9.5 ***Illustration of Some Available Commercial Equipment***
(Source: Niemira, Brendan A. Food Safety and Intervention Technologies Research Unit; US Department of Agriculture, Agricultural Research Service)

Production of the Non-Thermal Plasma

The author defines that "Non-Thermal Plasma" is high pressure (typically 1 atmospheric pressure) non-equilibrium plasma meaning the electron temperature is generally much higher than other particles, such as molecules (and sometimes ions). For semiconductor production process, similar non-equilibrium plasma is widely used and name as the cold plasma which is usually at the low pressure (several mTorr ~ Torr). Compared with that cold plasma, the electron temperature and ionization rate are much lower in the Non-Thermal Plasma. The name, "Non-Thermal Plasma" is often used against the arc which sometimes name as the Hot Plasma or Thermal Plasma. The arc is generated by DC, AC (commercial electric power frequency, 50 or 60 Hz), or microwave and is sometimes called the plasma torch where the gas temperature is from a few thousands to more than ten thousands Kelvin degrees.

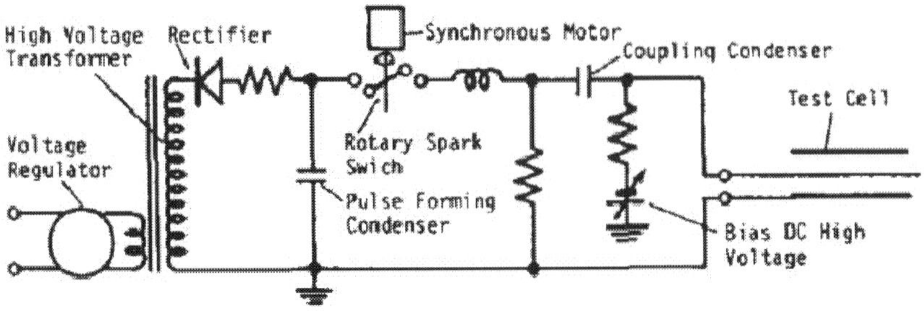

Figure 9.5 ***A High Voltage Power Supply With a Rotary Spark Gap***
(Source: ICESP X – Australia 2006; Paper 1A1 – Masuda Lecture: Atmospheric pressure non-thermal plasma decomposition of gaseous air contaminants and that diagnosis; Tetsuji Oda)

Pulse Discharge Plasma

For the ESP, the pulse width of the excitation voltage is typically microseconds, but the sub microseconds pulse is necessary for PPCP because the high pulse voltage is necessary to use the chemical reactions and that high voltage can easily cause the arc. Therefore the switching device must operate very quickly. The typical switching device for that is the discharge gap switch. For a long life switch, late previous researchers used a rotary spark gap switch. For a low frequency pulse or a small capacity test power supply, a self-triggered or pulse-triggered spark gap switch is also used as the switch device. There are several circuits to realize such short pulse. In Figure 9.6 the parallel resistance with the load determines the voltage tail time constant and the inductance adjusts the rising time. Below 100 kV, a stacked-semiconductor switch can realize ns switching performance, but is very expensive.

Recent IGBT (Insulated Gate Bipolar Transistor) is not yet usable at so high voltage, only 4.5 kV etc and cannot switch the high voltage directly. Another technology is the Blumlein pulse generator and the pulse compression technology. In the place of the capacitance, Blumlein cable can realize a very good waveform (sharp rising and falling voltage and a rather flat high voltage) and many researchers use that line.

Another technology to make good short high voltage pulse is the pulse compression technology. A typical one is the magnetic pulse compression which uses the magnetic saturation effect (change of impedance). That technology is widely used for the laser power supply. The demerit of that is that the pulse peak voltage is not variable. Toshiba proposed to use that device for the removal of the dioxin in the incinerator flue gas in the ESP.

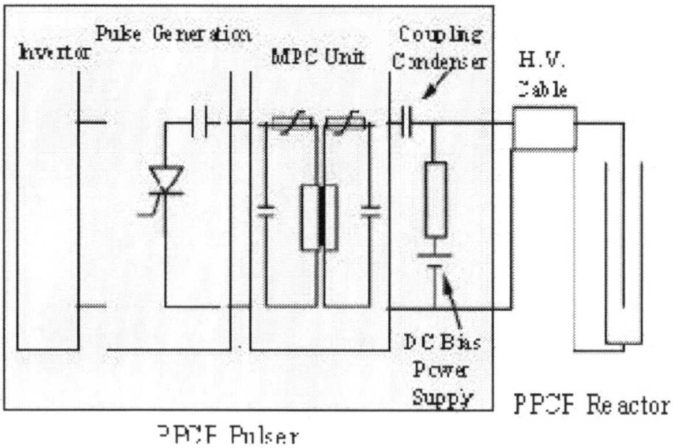

Figure 9.6 ***PPCP Power Supply***
(Source: ICESP X – Australia 2006 Paper 1A1 – Masuda Lecture; Atmospheric pressure nonthermal plasma decomposition of gaseous air contaminants and that diagnosis; Tetsuji Oda)

The pulse plasma is generally used in a large scale and the typical discharge is consisting of the spiral electrode and the center line electrode (discharge electrode). Figure 9.6 is one PPCP (pulse generator circuit 5) and Figure 9.7 is a photo of the pulse generator using the magnetic pulse compression. Some researchers use a disc-like discharge electrode. Previous researchers observed the streamer development around the disc electrode and found that the fast rising time is very important to generate a uniform and strong discharge. The authors also use a coaxial Marx pulse generator with a very short rising time (10 ns) and a short pulse width (60 ns) where the maximum output voltage is 500 kV in the matching condition. That generator has 12 SF6 spark gap switches to be triggered simultaneously and all capacitances are stored in a coaxial oil tank.

Figure 9.7 ***A photo of PPCP Power Supply***
(Supplied by Masuda Research Institute. Source: ICESP X – Australia 2006; Paper 1A1 – Masuda Lecture: Atmospheric pressure non-thermal plasma decomposition of gaseous air contaminants and that diagnosis; Tetsuji Oda)

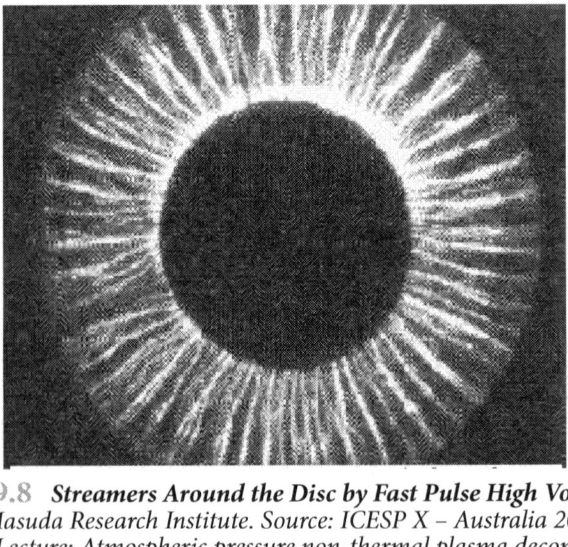

Figure 9.8 ***Streamers Around the Disc by Fast Pulse High Voltage***
(Supplied by Masuda Research Institute. Source: ICESP X – Australia 2006; Paper 1A1 – Masuda Lecture: Atmospheric pressure non-thermal plasma decomposition of gaseous air contaminants and that diagnosis; Tetsuji Oda)

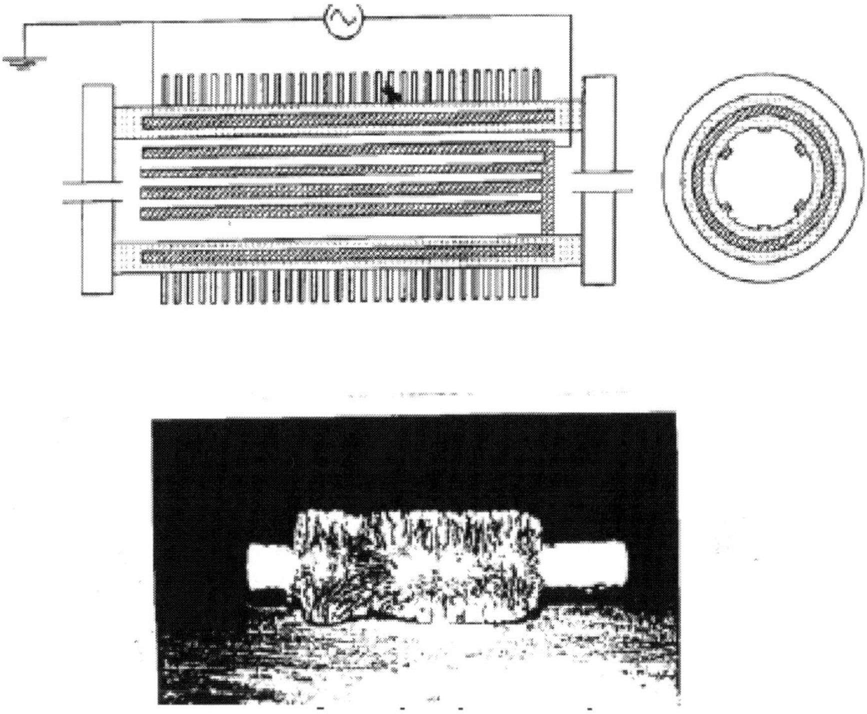

Figure 9.9 ***Surface Discharge Plasma Reactor (ceramic ozonizer)***
(Source: ICESP X – Australia 2006; Paper 1A1 – Masuda Lecture: Atmospheric pressure nonthermal plasma decomposition of gaseous air contaminants and that diagnosis; Tetsuji Oda)

Surface Discharge Plasma

As the pulse discharge plasma is excellent to generate the good non-thermal plasma but expensive and too much noisy. The surface discharge mode (including the space barrier discharge) is much easy to handle and quiet not only mechanically but also electrically. Still now, most typical application of the barrier discharge is the generation of the ozone.

Previous researchers used SPCP also for not only the DeNOx but also improving bad smell. If the AC (50 Hz–40 kHz) high voltage is applied between two electrodes, the discharge occurs on the inner ceramic surface which is named as the surface discharge. For the ozonizer, kHz high frequency high voltage is used but for the non-thermal plasma generation, 50 or 60 Hz low frequency is rather used because of the energy efficiency. The author uses a several types of the reactor as shown in Fig. 10.6. In many cases, the (dielectric) barrier discharge known as the ozone generation is very effective in generating the active non-thermal plasma. As the discharge electrode, the bolt type is very effective because of that sharp and uniform edge. However, the spacing factor of those cylindrical reactors is not good (dead space exists if they are bundled) and the plate type plasma reactor is also developed.

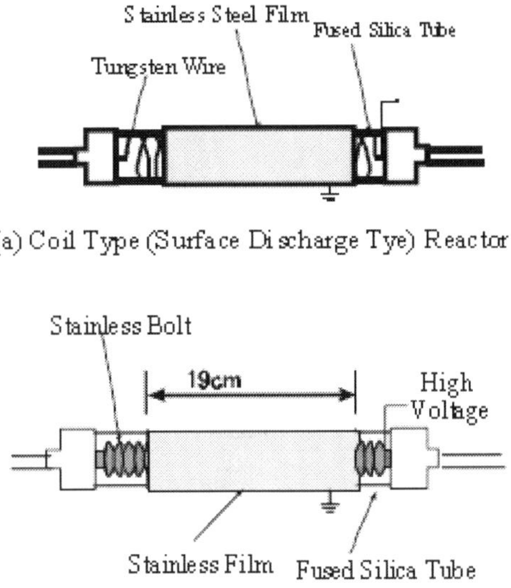

Figure 9.10 ***Non-Thermal Plasma Reactors***
(Source: ICESP X – Australia 2006; Paper 1A1 – Masuda Lecture: Atmospheric pressure non-thermal plasma decomposition of gaseous air contaminants and that diagnosis; Tetsuji Oda)

References

Niemira, Brendan A. *Non-thermal Processing with Plasma Technologies*. Food Safety and Intervention Technologies Research Unit. US Department of Agriculture, Agricultural Research Service. Print.

Moman, Raja M.; and Najmaldeen, Hmeda. "The bactericidal efficacy of cold atmospheric plasma technology on some bacterial strains." *Egypt. Acad. J. biolog. Sci.*, 2(2) (2010) : 43-47. G. Microbiology; ISSN: 2090–0872;. Print.

Oda, Tetsuji. Masuda Lecture: "Atmospheric pressure non-thermal plasma decomposition of gaseous air contaminants and that diagnosis." ICESP X – Australia 2006. Paper 1A1. Dept. Electr. Eng., the University of Tokyo. Print.

Guillard, V.; Mauricio-Iglesias, M.; and Gontard, N. "Effect of Novel Food Processing Methods on Packaging:; Structure, Composition, and Migration Properties". http://www.tandfonline.com/loi/bfsn20. Web.

Pedrow, P.; Wemlinger, E.; Alhamarneh, I. *Atmospheric pressure cold plasma processing of bioactive packaging applied directly to fresh fruits and vegetables*; Washington State University.

Chapter 10

Overarching Principles: Kinetics and Pathogens of Concern for All Alternative Food Processing Technologies

Kinetic Parameters for Inactivation of Microbial Populations

Models and Parameters

Kinetic parameters and models are used for the development of food preservation processes to ensure safety. They also provide the tools to compare the impact of different process technologies on reduction of microbial populations. The parameters used to analyze and report the reduction of a microbial population as a function of process parameters include empirical coefficients experimentally determined from microbial reduction kinetics, as well as constants from expressions based on chemical reaction kinetics. The purpose of this section is to present the models and kinetic parameters used to present and compare microbial inactivation data from thermal, pressure and electromagnetic processes.

Rate Constants

The traditional approach to describing changes in microbial populations as a function of time has used the survivor curve equation:

$$\log [N / N_0] = -t / D \qquad \text{(Eq. 10.1)}$$

Where:
N = microbial population at any time, t
N_0 = initial microbial population
D = decimal reduction time, or time required for a 1-log cycle reduction in the microbial population.

The corresponding model from chemical reaction kinetics is the first-order kinetic model:

$$dN / dt = -kN \qquad \text{(Eq. 10.2)}$$

Where:
k = reaction rate constant (first-order), or the slope of the natural logarithm of survivors in contrast to time for the microbial population.

Equation (10.2) can be integrated to obtain a more familiar expression for the reduction of microbial populations:

$$\ln [N / N_0] = -\,k\,t \qquad \text{(Eq. 10.3)}$$

By comparing Eq. (10.1) and (10.3), the relationship between the decimal reduction time and the first-order reaction rate constant is:

$$k = 2.303 / D \qquad \text{(Eq. 10.4)}$$

The primary parameters (D-value or k) would describe the microbial population reduction at a constant and defined temperature, pressure and/or electric field. The inherent assumption in the use of these models (and the corresponding parameters) is that the reduction in microbial population is described by the first-order reaction model. Alternative models are being developed to explain microbial inactivation kinetics when the linearity of the data is questionable.

If there is evidence of a different reaction model, different parameters need to be identified and used for process development and prediction purposes. Only a limited amount of the published data on microbial inactivation has been analyzed using the reaction rate model to quantify first-order rate constants (k). On the other hand, most published data on changes of food quality attributes have been presented as reaction rate constants (k). As indicated by the relationship between D-value and k, published data can be easily transformed.

Temperature Coefficients

Traditionally, the influence of temperature on microbial population inactivation rates has been expressed in terms of the thermal resistance constant (z-value) using the following model:

$$\log [D / D_R] = -(T - T_R) / z \qquad \text{(Eq. 10.5)}$$

The thermal resistance constant z(T) is the temperature increase needed to accomplish a 1-log cycle reduction in the D-value. The reference decimal reduction time (D_R) is the magnitude at a reference temperature (T_R) within the range of temperatures used to generate experimental data. Microbial populations with higher resistance to temperature change are described by larger z(T).

The most evident examples are the larger z(T) for spores compared to the ones for vegetative cells. An alternative model for describing the influence of temperature on microbial population reduction rates is the Arrhenius equation. The model illustrates the influence of temperature on the reaction rate constant (k), as follows:

$$k = k_0 \exp [-E / R\, T_A] \qquad \text{(Eq. 10.6)}$$

Where:
k_0 = Arrhenius Constant
E = Activation Energy Constant
T_A = Absolute Temperature
R = Universal Gas Constant

Based on the Arrhenius model (Eq. 10.6), the slope of ln (k) in contrast to $1/T_A$ plot determines the temperature coefficient E (activation energy constant). The activation energy constant describes the influence of temperature on the magnitude of the first-order reaction rate constant (k).

When the thermal resistance model and the Arrhenius model are applied to microbial population reduction rate data over the same temperature range, a relationship between the 2 coefficients [z(T) and E] is evident. By comparison of Eq. (10.5) and (10.6), the following relationship can be obtained:

$$\mathbf{E = 2.303\ R\ TA2\ /\ z} \qquad \text{(Eq. 10.7)}$$

The temperature used in Eq. (10.7) should be selected as a midpoint in the range of temperatures used to generate the original experimental data. Equation (10.7) does suggest that the relationship between the two temperature coefficients [E and z(T)] depends on temperature. The magnitudes of the two coefficients, however, are significantly different, and any influence of temperature is negligible as long as the temperature reference is within the range used for data collection. The use of the coefficients [z(T) or E] should be limited to the range of temperatures used to obtain experimental D-values. The z(T) should only be used with a defined reference temperature as emphasized by Datta.

The use of the first-order models and the corresponding models for temperature influence must be applied within the limits of data used to generate the parameters within the expressions. The estimation of the kinetic parameters from the appropriate model requires careful attention to statistical limits created by the experimental data. Several authors, including Arabshahi and Lund and van Boekel, have demonstrated the influence of statistical parameters on the use of the prediction models.

Pressure Coefficients

There are only limited references to parameters used to describe the influence of pressure on the rate of microbial population reduction. Zook and others have used a parameter similar to the thermal resistance constant z(T), based on the following model:

$$\log[D / D_R] = -(P - P_R) / z \qquad \text{(Eq. 10.8)}$$

Where:
D_R = decimal reduction time at a reference pressure (P_R).

In this report, the pressure coefficient will be defined as:

z(P) = the pressure increase required to accomplish a 1-log cycle reduction in the decimal reduction time (D-value).

In order for pressure resistance constant z(P) to be meaningful, it is important to include a minimum of 3 D-values in the analysis of data. All D-values must be obtained at the same temperature and above the threshold pressure needed for the target microbial inactivation. The threshold pressure (or critical pressure) is the pressure below which microbial inactivation does not occur. An alternate model to describe the influence of pressure on microbial inactivation rates is based on the Eyring equation, as proposed by Weemaes and others. The model describes the reaction rate constants (k) as follows:

$$\ln (k) = \ln(k_R) - [V(P - P_R)/RT_A] \qquad \text{(Eq. 10.9)}$$

Where:
k_R = reaction rate constant at reference pressure (P_R)
V = activation volume constant
P = pressure
T_A = absolute temperature

The activation volume constant (V) is the pressure coefficient obtained from the slope of the ln (k) in contrast to ($P - P_R$) plot. The magnitude of V increases as the slope of the plot increases. When the rate of microbial inactivation increases significantly with small changes in pressure, the magnitude of the V will be larger.

Alternatively, smaller values of V describe microbial populations with inactivation rates that would change less when pressure changes. As suggested when describing z(P) values, it is important for all reaction rate constants (k) used in the analysis to be measured at the same temperature. For the activation volume constant (V) to be useful and meaningful, the k constant should be measured at pressures above the threshold pressure needed to inactivate the target microbial population.

Electric Field Coefficients

As in the case of pressure processes, when microbial populations are exposed to pulsed electric fields (PEF), the electric field intensity applied should be above the threshold electric field intensity, the critical electric field intensity for the target microorganism. A model similar to those for temperature and pressure can be used to describe the influence of electric field intensity on the rate of microbial population reduction. The proposed model would be:

$$\log [D / D_R] = -(E - E_R) / z \qquad \text{(Eq. 10.10)}$$

Where:
D_R = decimal reduction time at a reference electric field intensity (E_R).

The electric field coefficient in this model is defined as:

z(E) = the increase in electric field intensity (E) required to reduce the decimal reduction time (D) by 1-log cycle at a specific temperature and pressure.

All D-values used in this type of analysis should be acquired at the same temperature and pressure. A minimum of 3 D-values should be obtained for the data analysis. An alternative model for describing the influence of electric field intensity on the survival of a microbial population was proposed by Peleg. The model is based on the Fermi equation and can be expressed as:

$$\mathbf{N / N_o = 1/\{1 + exp[(E - E_d) / K]\}} \qquad \text{(Eq. 10.11)}$$

Where:

E_d = the electric field intensity when microbial population has been reduced by 50%.

K = a coefficient with magnitude based on the slope of the survivor curve obtained at several levels of electric field intensity.

This model has been applied to survivor data for several different microbial populations to generate typical magnitudes of the coefficient (K). Larger magnitudes of the coefficient would suggest a higher resistance to changes in electric field intensity.

A similar model has been proposed and used by Hulsheger and others and applied by Jeyamkondan and others. The model describes the survivor number as a function of electric-field strength and treatment time:

$$\mathbf{N / N_o = \{ t / t_c \}[- (E - E_c) / K]} \qquad \text{(Eq. 10.12)}$$

Where:
t = treatment time
t_c = critical treatment time or treatment time below which no inactivation of microorganisms occurs
E_c = critical electric field strength or electric field strength below which no inactivation of target microorganism occurs
K = specific rate constant

The model proposed by Hulsheger and others is similar to Eq. (10.11), but accounts for exposure time at a given electric field intensity. The coefficient (K) has a similar relation to electric field intensity as in Eq. (10.11) and the relative magnitudes should be interpreted in the same manner.

Kinetic Parameters for Inactivation of Microbial Pathogens

The purpose of this section is to provide an overview and discussion on the kinetic data of microbial population inactivation. This section addresses the use of kinetic parameters for development of processes and the comparison of parameters obtained for various microorganisms, including a discussion on the limitations of the parameters. Finally, the research needs will be addressed, with specific attention to recommendations on experimental approaches to be considered in the future.

Kinetic parameters describing the inactivation of microbial pathogens are presented in Tables 10-1a, 1b and 1c and are a summary of parameters presented in other sections of this report. The intent of the summary is to provide an overview and a comparison of the kinetic parameter magnitudes for the various microorganisms for each process technology.

The parameters defined before (D-value and z(T), z(P), z(E), E, k, K and V) have been calculated from data previously reported and using the models here for thermal, pressure and PEF technologies. The parameters for thermal treatment also apply to microwave energy and electrical resistance (ohmic) processes, as well as any other technology where temperature is the primary factor in reduction of the microbial population.

Likewise, the parameters for pressure or PEF treatments should apply to any process where pressure or electricity is the primary critical factor in reducing microbial populations. It must be noted that, given the scarcity of data, these are estimated parameters and there is an imminent need for more research in this area. Although this report contains references to several other technologies, the quantity of data describing the influence of the treatment on reduction of microbial populations is insufficient at this time.

Like in most of the published literature, in this report data have been analyzed assuming that the reduction in microbial populations follows a linear first-order model, with the exception of the PEF parameters that will be discussed in Section Pulsed electric Field Processes. The potential of non-linear inactivation data or the use of alternative models cannot be ignored.

Because there is currently insufficient information on alternative models to allow the type of comparisons being considered in this portion of the report, these issues will be discussed when describing the specific technologies. The use of consistent parameters for all preservation technologies should improve the efficiency of future investigations and encourage uniformity in the methodologies for establishment of minimum process requirements.

Process Development

The parameters presented in Tables 10-1a, 1b and 1c parallel the traditional parameters used for development of thermal preservation processes. The basic model for process development is based on the survivor curve Eq. (10.1) or (10.2):

$$\mathbf{F = -D \log [N_0 / N] = D \log [N / N_0]} \quad \text{(Eq. 10.13)}$$

or:

$$\mathbf{F = -\ln [N_0 / N] / k = \ln [N/N_0] / k} \quad \text{(Eq. 10.14)}$$

Where

F = the total time required to reduce the microbial population by a specified magnitude needed to ensure product safety, under the conditions defined by D-value or k.

The basic model assumes a linear first-order relationship between microbial population and time. Currently, there is a lack of historical evidence to support alternative models; however, there is considerable discussion about the appropriateness of using a first-order model to describe the reduction in microbial population for all preservation technologies. For example, models for PEF technology as presented in Eq. (10.11) and (10.12) should continue to be evaluated, but at this time, input parameters for these models are limited.

Inactivation Data and Parameters

Limitations of the Calculated Parameters

A few limitations need to be considered when interpreting the parameters presented in Tables 10-1a, 1b and 1c. Care should be taken when they are intended to be used as tools to develop processes, to compare the resistance of different microbial populations, or to identify appropriate surrogate microorganisms. As illustrated in this book, the kinetic parameters for microbial populations exposed to thermal treatments have been assembled over a significant period of time.

Over time, the published literature has included kinetic parameters needed to respond to most process, product and microbial situations. The parameters provide a sound basis to develop processes for the microwave energy and electrical resistance (ohmic) technologies. In addition, the available parameters provide a sound basis to compare different microbial populations and the influence of different product environments on the parameter magnitudes.

The key issue for these electrothermal treatments is the lack of conclusive evidence on the existence of non-thermal effects influencing the reduction in microbial populations. It is believed, however, that those effects would add an extra factor of safety to the preservation process (See microwave, and ohmic, and inductive Heating chapters).

In general, the data used to determine the D-values (and k-values) for pressure processes appear to be adequate. The limitations to these data are primarily associated with temperature control during pressure treatments. In addition, when temperature changes have been reported, the influence on the kinetic parameters has not been analyzed. The evidence suggesting a synergistic impact of pressure and temperature is too limited for use in process evaluation.

The most serious deficiency in pressure process kinetics is that most of the parameters (D and k) have been measured at a single pressure.

Only 4 studies have used 3 to 5 pressure levels, while controlling all other factors influencing the parameters. The results from these studies are adequate to evaluate the pressure coefficient [z(P)] and/or activation volume (V). With exception of the 4 publications cited above, the estimated parameters are limited by the number of pressure magnitudes used, the lack of temperature control and the lack of multiple data for the same microorganism and/or product/substrate. By overcoming these limitations, parameters from future investigations will meet the needs of process development and product/microorganism comparisons.

The data available on the influence of PEF on microbial populations have many limitations. As will be indicated during the discussion of parameters in Table 10-1c, the kinetic parameters (D-value or k) are based on two points on the survivor curve, the initial population and the final population. It should be recognized that the values of parameters in Table 10-1c were not based on linear regression analysis. In addition, temperature controls and collection of multiple data points at the same temperature level are lacking.

At this time, no single report has measured the inactivation of microbial populations at several levels of electric field strength, leading to the quantification of the PEF coefficient z(E). Although 3 such coefficients are presented in Table 10-1c, these coefficients have been estimated based on kinetic parameters reported in separate investigations and must be used with these limitations in mind. There are no published reports that evaluate the potential for a synergistic influence of electric field strength and temperature. There are only 2 reports with kinetic parameters based on Eq. (10.11) and (10.12) and these reports provide limited parameters on microorganisms of food safety concern. They do not include any of the microorganisms of food safety concern.

Thermal Processes

The literature provides an impressive array of kinetic parameters to be used in the development of thermal processes. In addition to data and parameters on inactivation of microbial populations, Table 10-1a includes additional information on the medium used and specific experimental conditions (that is, temperature) when available. The time parameters are the decimal reduction time (D-value) and the corresponding rate constant (k). The temperature coefficients include the thermal resistance constant [z(T)] and the activation energy constant (E).

The kinetic parameters calculated for the thermal inactivation of microbial pathogens in Tables 10-1a, 1b and 1c should be considered when using any process technology where temperature is the primary mode of microbial inactivation. The most promising alternative thermal processes to reduce pathogenic microbial populations are microwave energy and electrical resistance (ohmic), which are included in this report.

As suggested in the chapters on microwave, and ohmic, and inductive heating, it is assumed that the direct influence of microwave energy or electrical resistance on microorganisms is negligible. Therefore, thermal kinetic parameters should be considered for the abovementioned electrothermal processes (ohmic, inductive and microwave heating).

Kinetic parameters for vegetative cells of *Salmonella* serovars, pathogenic *Escherichia coli, Yersinia enterocolitica*, pathogenic *Vibrio spp., Aeromonas hydrophila, Campylobacter jejuni, Listeria monocytogenes* and *Staphylococcus aureus* are presented in Table 10-1a. In general, the D-values are relatively small and the k-values are relatively large for the vegetative microorganisms normally targeted in pasteurization or other mild thermal processes. Other than the abnormally high D-values (low k-values) for *Salmonella* pathogens in milk chocolate, *Salmonella Typhimurium* and *L. monocyctogenes* are the most thermally resistant vegetative microorganisms.

The largest D-value for *Salmonella Typhimurium* is 18.3 min (k= 0.126/min) at 55 °C. For *L. monocyctogenes*, the largest D-value is 16.7 min (k = 0.14/min) at 6 °C. The largest D-values for *E. coli* are 6.6 min (55 °C) for O111:B4 and 6.4 min (57 °C) for O157:H7. Based on limited data for O157:H7 in ground beef, a z(T) of 5.3 °C has been estimated.

Other significant magnitudes for D-values include 6.6 min for *A. hydrophila* at 48 °C and 16.7 min for *L. monocytogenes* in cured ground beef at 60 °C. A z(T) of 5.56 °C for *L. monocytogenes* in milk has been estimated, based on published data. In general, the thermal resistance constants z(T) for the vegetative microorganisms fall in the range between 4 and 7.7 °C. This range includes a z(T) of 5.3 °C for *E. coli* O157:H7 in ground beef and of 5.56 °C for *L. monocytogenes* in milk, both estimated from limited data presented in the references cited. The larger z(T) presented include 12.4 to 25 °C for *Vibrio* species (in fish products) and 17.7 to 18.9 °C for *Salmonella serovars* (in milk chocolate). These abnormally high z(T) for vegetative microorganisms should be noted for these products and may be specifically associated with them.

S. aureus, a vegetative microorganism that produces a heat-stable toxin, has D-values similar to other vegetative populations. The z(T) of 9.5 °C is relatively high and must be considered when developing processes for situations where *S. aureus* could present a health hazard. The largest D-value (smallest k-value) reported at 110 °C for toxin-producing, sporeforming microorganisms is 12.42 min (0.185/min) for *Clostridium botulinum* proteolytic Type B spores in pureed peas. Most other D-values are in the more typical range of 1 to 3 min for spore-forming microorganisms.

Other values to be noted are the D-value of 36.2 min (k = 0.064/min) for *Bacillus cereus* spores at 95 °C and 100 min (k = 0.023/min) for *C. botulinum* non-proteolytic Type E spores at 70 °C. When expressed at 110 °C, these D-values become 1.18 min for the *B. cereus* spores and less than 1 sec for the Type E spores. Data for *Bacillus subtilis* spores have been included in Table 10-1a to illustrate the influence of ohmic heating on inactivation kinetics.

These data were reported by Cho and others (1999) and indicate that the reduction in D-value (higher k-value) and the increase in z(T) (lower E) when using ohmic heating are statistically significant. These results suggest an independent and additional inactivation mechanism due to the electric current during the ohmic heating. The overall influence of these non-thermal effects, however, is not sufficient to consider the use of alternate kinetic parameters for development of ohmic heating processes.

These authors have demonstrated that a 2-stage process involving ohmic heating, interrupted by a 20-min incubation, resulted in enhanced inactivation of *B. subtilis* spores. This increase in inactivation has been attributed to the positive influence of electric treatment on spore germination.

Separate data for microwave heating are not included in this section. The non-thermal effects of microwave processes on microbial inactivation have not been confirmed and appear to be of insufficient magnitude to be considered during development of processes.

Pressure Processes

For processes involving the use of pressure for reduction of microbial populations, the F-value is the time the product needs to be exposed to the specified pressure and other conditions (that is, temperature) to accomplish the recommended amount of inactivation. Since the application of most of the pressure technologies involves instantaneous adjustment to the process pressure, the use of the basic model is straightforward.

The pressure coefficients [z(P) or V] provide users with the flexibility to select the most appropriate pressure for the specific application. For pulsed-pressure technologies, the model would need to incorporate the influence of time and incremental pressure. In this case, the estimation of kinetic parameters will require the measurement of other variables.

The kinetic parameters for inactivation of microbial populations due to pressure are presented in Table 10-1b. The time parameters, decimal reduction time (D-value) and first-order rate constant (k), were calculated based on the reduction in microbial population at a constant pressure. The pressure coefficients are z(P) and the activation volume constant (V), as defined in Chapter 1, and indicate influence of pressure on the rate of inactivation.

In most references cited, there are insufficient data to estimate these coefficients. Special consideration needs to be given to the combined use of pressure and temperature. Based on the current available information, the z(P) and z(T) parameters should be adequate for process development. The combined influence of pressure and temperature on inactivation kinetics for microbial populations has been investigated, although not extensively.

Published reports suggest a synergistic impact of pressure and temperature on inactivation rates, but additional investigations are needed. The independent influence of pressure on rates, as indicated by the z(P) or V parameters, needs to be clearly established. The influence of temperature can be quantified in several ways, but the optimum approach would be based on the dependence of z(P) or V on temperature. Although minimum pressure thresholds for microbial inactivation are not presented in this section, these parameters are discussed in the section on high pressure processing.

Several investigations on *Salmonella* indicate that decimal reduction times (D-value) range from 1.48 to 6 min (k = 0.348 to 1.556/min), with pressure having an obvious influence on the rate. Most of the studies have been conducted at ambient temperatures (20 to 25 °C). The D-values for *E. coli* are as high as 15 min (k = 0.154/min) at 300 MPa and 6 min (k = 0.384/min) at 600 MPa for O157:H7.

There are insufficient data to establish the influence of pressure or temperature and therefore z(P) or z(T) were not estimated. Pressure appears to have a significant influence on inactivation rates for populations of *S. aureus*, apparently one of the most pressure-resistant vegetative bacteria, as suggested by D-values of 7.14 min (k = 0.323/min) at 600 MPa compared to 150 min (k = 0.015) at 400 MPa. D-values reported for 500 MPa are lower than the ones for 400 MPa, but were measured in a different medium and may be influenced by temperature.

However, in comparable experiments, inactivation rates of selected strains of various *Listeria spp.* with, for example, D-values ranging from 1.48 min (k = 1.556/min) at 350 MPa to 15 min (k = 0.154/min) at 400 MPa were lower than the ones for *S. aureus*. These data were measured at ambient temperatures (20 to 25 °C). Recently, D-values of over 5 min were also reported for *L. monocytogenes* at 345 MPa and 25 °C (Alpas and others, 1999).

Comprehensive data on inactivation rates of *Clostridium sporogenes* spores were reported by Rovere and others (1996). These data indicate that D-values are 0.695 min (k = 3.314/min) at 800 MPa at 108 °C compared to 16.772 min (k = 0.136/min) at 600 MPa at 90 °C. The magnitudes of these D-values are similar to the D-value of 12 min at 680 MPa reported in a separate investigation, even though the latter was measured at ambient temperatures.

From the Rovere and others (1996) data, the influence of pressure on inactivation rate, z(P), were estimated to be 725 MPa at 93 °C,

962 MPa at 100 °C and 752 MPa at 108 °C. The inconsistent influence of temperature on z(P) may be associated with the limited range of temperatures and pressures used in the experimental investigation, as well as adequacy of temperature control during data collection.

Recent inactivation data for *C. botulinum* Type E Alaska and Type E beluga (Reddy and others, 1999) indicate that their D-values were in the same range as for *C. sporogenes*. The D-values for *C. botulinum* Type E Alaska were lower in crab meat than in a buffer. The D-values for *C. botulinum* Type A 62-A are generally higher than the values for *C. sporogenes*, even when considering the influence of temperature and pressure. The pressure coefficient z(P) for the Type A 62-A data was 1524 MPa. Surprisingly, this value was much higher than the z(P) values reported for *C. sporogenes*, even though data from *C. sporogenes* were recorded at lower temperatures.

An in-depth investigation of pressure inactivation of *Saccharomyces cerevisiae* in orange and apple juice has been reported by Zook and others. The calculated D-values were 10.81 min (k = 0.21/min) at 300 MPa, where temperatures have been maintained at levels between 34 and 43.4 °C. These D-values are slightly higher than the ones reported earlier by Parish and others (1998). For apple juice and orange juice z(P) were 115 MPa and 117 MPa, respectively. These values are much lower than those reported for *C. sporogenes* and *C. botulinum*. Since data for 5 different pressures have been reported by Zook and others (1999), the activation volumes (V) could be estimated to be 1.24 X 10-4 for orange juice and 1.37 X 10-4 m3/mole for apple juice.

In summary, the most pressure-resistant pathogenic vegetative cell populations appear to be those of *E. coli* O157:H8 with a D-value of 6 min (k= 0.384/min) at 600 MPa, and *S. aureus* with a D-value of 7.14 min (k = 0.323/min) at 600 MPa. The most pressure-resistant spores appear to be *C. sporogenes* with a D-value of 16.772 min (k = 0.138/min) at 600MPa (T = 90 °C) and *C. botulinum* Type A 62-A with a D-value of 6.7 min (k = 0.344/min) at 827 MPa (T = 75 °C). The pressure coefficient z(P) of 1524 MPa at 75 °C for *C. botulinum* Type A 62-A constitutes an additional indication of the pressure resistance of the spore populations. A recent report shows little if any inactivation after 30 min of *C. botulinum* 17B and Cap 9B exposure to 827 MPa at 75 °C (Larkin and Reddy 1999).

Pulsed Electric Field Processes

Currently, the majority of the kinetic parameters for the PEF technologies are in a form that fits the basic model [Eq. (10.13) or (10.14)]. Even with the limitations mentioned above, the use of the parameters and model to establish process time (F) would seem appropriate in the short term. Models, such as Eq. (10.11) or (10.12), provide desirable alternatives, but a great effort would be needed to evaluate them. The use of z(E) values provides the users with flexibility to select the optimum electric field strength for a given product and to evaluate the influence of other factors such as synergistic effects of electric field strength and temperature.

Adequate inactivation data for estimating the kinetic parameters for microbial populations exposed to PEF are scarce. The information presented in Table 10-1c compares decimal reduction times (D-value) and first-order rate constants (k), for experiments where electrical field strength (E) and initial temperature were mostly available. Three different PEF coefficients have been presented: the z(E), the specific rate constant (K) from the Hulsheger model and a similar constant (K) based on the analysis by Peleg.

It should be noted that the D-values (k-values) have been determined from measurements of microbial population reduction after 1 exposure time to a given electrical field strength. The parameters obtained should be considered with this limitation. Furthermore, there is no evidence that the survivor curve during exposure to a pulsed electric field is described by a first-order model.

The parameters are presented in this report to allow for more direct comparisons of the effectiveness of PEF in reducing different microbial populations, as well as to note the influence of the media on microbial inactivation. In addition, the D-values (k) provide a more direct approach to evaluating the influence of electric field strength on the rate of microbial population reduction. As will be emphasized later in this section of the report, there is a great need to better understand survivor curve shapes for microbial populations exposed to pulsed electric fields.

The results in Table 10-1c clearly indicate that the D-values are several orders of magnitude smaller than the same parameters for thermal or pressure processes. Assuming first-order kinetics through 6-12Ds, this suggests a significant advantage for PEF, when compared to the other technologies. This assumption may not be valid because inactivation of 99.9% of a cell population is frequently difficult to achieve.

Several investigations have reported data on reduction of *E. coli* populations exposed to PEF. The highest D-values are 4500 μs (k = 0.051 X 10-2 /μs) at 16 kV/cm and 17.8 μs (k = 12.94 X 10-2 /μs) at 70 kV/cm. Using a limited number of D-values, a z(E) of 41 kV/cm has been estimated. Note that this magnitude is based on less than ideal data, collected at temperatures ranging from 15 to 37 °C. The D-value of 4500 μs, at 16 kV/cm and 37 °C for *E. coli* would suggest that this microorganism is one of the more PEF resistant vegetative cell populations.

The investigations on the influence of PEF on *Salmonella* Dublin, *S. aureus* and *Zygosaccharomyces bailii* provide only limited amounts of data. The D-values for *S. aureus* are very similar to the magnitudes for *E. coli*, with values of 4000 to 6000 μs at relatively low electric field strength (16 kV/cm) and temperatures of 30-37 °C.

The data for the *Listeria spp.* indicate that D-values are as low as 18.8 μs at 50 kV/cm for *Listeria innocua* and as high as 540 μs at 20 kV/cm for *L. monocytogenes.* Since these data were measured at relatively low temperatures (10 to 50 °C), the parameters would indicate that *Listeria* is one the more resistant vegetative cell populations to a PEF treatment. The D-values of 50-60 μs (k = 3.84 to 4.61 X 10-2 /μs) at 50 kV/cm for *B. cereus* spores are higher than for other microbial populations at the same field strength and temperature.

Two D-values (17.5 to 26.3 μs) for *B. subtilis* spores at the same pressure and from 2 different investigations were considerably lower than the D-values for *B. cereus* spores. Using the D-values for *B. subtilis* spores at 3 different electrical field strengths and within an ambient temperature range, a z(E) of 15.5 kV/cm has been estimated. Unexpectedly, this magnitude is much lower than the one estimated for vegetative cell populations [that is, *E. coli* with a z(E) of 41 kV/cm]. These observations need more comprehensive investigation before any conclusions are reached.

Several investigations have reported data on inactivation of *S. cerevisiae* when exposed to PEF. Overall, the D-values vary significantly depending on the electric field strength and temperature. In general, the magnitudes are larger than *E. coli*, lower than *L. monocytogenes* and much less than *B. subtilis* spores. An z(E) of 17 kV/cm has been estimated from data reported for PEF treatments of *S. cerevisiae* in apple juice, much lower than the value estimated for *E. coli* (41 kV/cm) and similar to the one of *B. subtilis* spores (15.5 kV/cm).

The influence of electrical field strength (E) on the rate of microbial population inactivation may also be estimated from the coefficient (K). These parameters have been reported for a limited number of microbial populations. Among them, the populations with greater resistance to PEF would include *Escherichia spp.*, *Listeria spp.*, *Pseudomonas spp.* and *Klebsiella spp.* The coefficient z(E) was highest for *Escherichia spp.*, which was higher than the one for *B. subtilis* spores. Data are insufficient to make valid comparisons of the relative resistance for vegetative and spore populations to PEF.

In summary, the survivor data for microbial populations exposed to PEF are too limited to be used in reaching definite conclusions about the magnitude of the kinetic parameters. In addition, data are not adequate to calculate parameters to compare the relative resistance of various microbial populations to PEF. For instance, data based on the same field strength and temperature are lacking. In addition, only a few of the published reports provide information on the threshold field strengths needed to initiate inactivation.

Future Research Needs

This section focuses on the research needs associated with kinetic parameters to be used for development of food preservation processes to ensure safety. For several technologies discussed in this report, the data necessary to estimate kinetic parameters are lacking. If these technologies are to evolve to industrial applications, kinetic data must be collected in the future.

The following is a list of research areas that need further investigation:

- *Evaluation of the adequacy of a linear first-order survivor curve. Although there is evidence of various types of deviation from the historical model, a universally accepted alternative has not evolved. Future research on an appropriate model would be beneficial to all preservation technologies.*

- *Investigation on the influence of pressure on reduction of microbial populations using the proper experimental design (statistically valid, collection of data at different pressures and control of temperature and product), so that z(P) and/or activation volumes (V) are quantified. These investigations should also evaluate synergistic effects between pressure and temperature.*

- *Research on developing an experimental protocol for obtaining statistically reliable kinetic parameters to describe survivor curves for microbial populations exposed to PEF. These studies should incorporate multiple levels of electric field intensity, as well as the potential for synergy with temperature.*

- *Further research on the PEF microbial inactivation models presented as Eq. (10.11) or (10.12). The investigations need to provide reliable kinetic parameters for these models and for the microbial population of interest in food safety.*

Table 10-1a ***Kinetic Parameters for Inactivation of Microbial Population for Thermal Processes***

Microorganism	Substrate	Time Parameter		Temperature Coefficient		Temperature	Other	References
		(D)	(k)	Z(T)	(E)			
		(min)	(1/min)	(°C)	(kJ/mole)	(°C)		
Vegetative Cells								
Salmonella serovars	Milk	0.018-0.56	4.113-127.9	4.4-5.6	392-499	65.6		ICMSF (1996)
S.Senftenberg	Various Foods	0.56-1.11	2.075-4.113	4.4-5.6	392-499	65.5		ICMSF (1996)
S.Typhimurium	TBS + 10-42%MS	4.7 - 18.3	0.126-0.49	4.5-4.6	448-458	55		ICMSF (1996)
S. Senftenberg	Milk Chocolate	276 - 480	0.005-0.008	18.9	120	70-71		ICMSF (1996)
S.Typhimurium	Milk Chocolate	396 - 1050	0.002-0.006	17.7	128	70-71		ICMSF (1996)
S.Typhimurium	Ground Beef	2.13 - 2.67	0.86-1.08			57		ICMSF (1996)
S.Eastbourne	Milk Chocolate	270	0.0085			71		ICMSF (1996)
Escherichia coli ATCC	Dairy Products	1.3-5.1	0.45-1.77			57.2		ICMSF (1996)
E. coli O111:B4	Skim/ Whole Milk	5.5-6.6	0.35-0.42			55		ICMSF (1996)
E. coli O157:H7	Ground Beef	4.1-6.4	0.36-0.56			57.2		Line and et al. (1991)
E. coli O157:H8	Ground Beef	0.26-0.47	4.9-8.86	5.3	401	62.8		Line and et al. (1991)
Yersinia enterocolitica	Milk	0.067-0.51	4.52-34.4	4-5.78	367-530	60		ICMS (1996)
Vibrio parahaemolyticus	Fish Homogenate	10 - 16	0.144-1.05	5.6-12.4	159-352	48		ICMSF (1996)
V. parahaemolyticus	Clam/Crab	0.02-2.5	0.92-115	5.6-12.4	166-368	55		ICMSF (1996)
V. cholerae	Crab/Oyst	0.35-2.65	0.87-6.58	17-21	101-125	60		ICMSF (1996)
Aeromonas hydrophila	Milk	2.2-6.6	0.35-1.05	5.2-7.7	256-379	48		ICMSF (1996)
Campylobacter jejuni	Skim Milk	0.74 - 1.0	2.3 - 3.11			55		ICMSF (1996)
C. jejuni	Beef/Lamb/ Chicken	0.62 - 2.25	1.0 - 3.72			55-56		ICMSF (1996)

Listeria monocytogenes	Milk	0.22 - 0.58	3.97 - 10.47	5.5	386	63.3		ICMSF (1996)
L. monocytogenes	Meat Products	1.6 - 16.7	0.14 - 1.44			60		ICMSF (1996)
Staphylococcus aureus	Milk	0.9	2.56	9.5	224	60		ICMSF (1996)
S. aureus	Meat Macerate	6	0.384			60	+500 ppm nitrite	ICMSF (1996)
S. aureus	Pasta	3	0.768			60	$a_w = 0.92$	ICMSF (1996)
S. aureus	Pasta	40	0.0576				T= 60 °C, $a_w = 0.8$	ICMSF (1996)
S. aureus	Phosphate Buffer	2.5	0.921			60	pH = 6.5	ICMSF (1996)
Spores								
Bacillus cereus	Various	1.5 - 36.2	0.064 - 1.535	6.7 -10.1		95		ICMSF (1996)
Bacillus cereus	Various	1.5 - 36.2	0.064 - 1.535	6.7 -10.1		95		ICMSF (1996)
Clostridium perfringens	Phosphate Buffer	0.015 - 8.7	0.265 - 15.35			90	pH = 7.0	ICMSF (1996)
C. perfringens	Phosphate Buffer	3.15	0.731			104.4	pH = 7.0	ICMSF (1996)
C. perfringens	Beef Gravy	6.6	0.349			104.4	pH = 7.0	ICMSF (1996)
Clostridium botulinum 62A	Vegetable Products	0.61 - 2.48	0.929 - 3.775	7.5 11.6		110		ICMSF (1996)
C. botulinum 62A	Phosphate Buffer	0.88 - 1.9	1.212 - 2.617	7.6 - 10		110	pH = 7.0	ICMSF(1996)
C. botulinum 62A	Distilled Water	1.79	1.287	8.5		110		ICMSF (1996)
C. botulinum B	Phosphate buffer	1.19 - 2.0	1.152 - 1.935	7.7 - 11.3		110	pH = 7.0	ICMSF (1996)
C. botulinum B	Vegetable products	0.49 - 12.42	0.185 - 4.7	7.4 - 10.8		110		ICMSF (1996)
C. botulinum E	Seafood	6.8 - 13	0.177 - 0.339	9.78		74		ICMSF (1996)
C. botulinum E	Oyster homogenate	72 - 100	0.023 - 0.32	6.8 - 7.5		70		ICMSF (1996)
Bacillus subtilis	0.1% NaCl	32.8	0.0702	8.74	293	88	Conventional	Cho and others (1999)
B. subtilis	0.1% NaCl	30.2	0.0763	9.16	282	88	Ohmic	Cho and others (1999)

Table 10-1b ***Kinetic Parameters for Inactivation of Microbial Population for Pressure Processes***

Microorganism	Substrate	Time Parameter		Pressure Coefficient		Pressure	Other	References
		(D)	(k)	[z(P)]	(V)			
		(min)	(1/ min)	(MPa)	(m^3 / mole)	(MPa)		
Vegetative Cells								
Campylobacter		< 2.5	>0.92			300		Smelt and Hellemons (1998)
Salmonella Senftenberg	Buffer	6	0.384			345	T= 23°C	Metrick and others (1989)
S. Senftenberg		5	0.461			300		Smelt and Hellemons (1998)
Salmonella Enteritidis	Meat	3	0.768			450		Patterson and et al. (1995)
Salmonella Typhimurium	Milk	3	0.768			350		Patterson and others (1995)
S. Typhimurium	Meat	1.48	1.556			414	T=25 °C	Ananth and others (1998)
S. Typhimurium		0.6	3.838			345	T=50 °C	Kalchayanand et al. (1998)
Yersinia enterocolitica	Milk	3	0.768			275		Patterson and others (1995)
Escherichia coli		7.5 - 15	0.154 - 0.307			300		Smelt and Hellemons (1998)
E. coli	Milk	1	2.303			400	T=50 °C	Gervilla et al. (1997b)
E. coli	Meat	2.5	0.92			400		Patterson and Kilpatrick (1998)
E. coli	Milk	1	2.303			450	T=25 °C	Gervilla and others (1997a)
E. coli	Buffer	3	0.768			700		Patterson and others (1995)
E. coli O157:H7	Milk	3	0.768			400	T=50 °C	Patterson and Kilpatrick (1998)
E. coli O157:H8		6	0.384			600		Smelt and Hellemons (1998)
E. coli O157:H7		0.7	3.29			345	T=50 °C	Kalchayanand et al. (1998)
Staphylococcus aureus		150	0.015			400		Smelt and Hellemons (1998)

S. aureus	Milk	2.5	0.92			500	T = 50 °C	Patterson and Kilpatrick (1998)
S. aureus	Meat	3	0.768			500	T = 50 °C	Patterson and Kilpatrick (1998)
S. aureus		7.9	0.292			500		Smelt and Hellemons (1998)
S. aureus		7.14	0.323			600		Smelt and Hellemons (1998)
S. aureus	Buffer	3	0.768			700		Patterson and others (1995)
S. aureus 582		0.6	3.838			345	T = 50 °C	Kalchayanand et al.(1998)
L. monocytogenes		1.48 - 13.3	0.173 - 1.556			350	101 strains	Smelt and Hellemons (1998)
L. monocytogenes	Milk	3	0.768			375		Patterson and others (1995)
L. monocytogenes	Meat	2.17	1.061			414	T = 25 °C	Ananth and others (1998)
L. monocytogenes ScottA	Meat	3.5	0.658			400	T=Amb.	Mussa and others (1999)
L. monocytogenes		5.0 - 15	0.154 - 0.461			400		Smelt and Hellemons(1998)
L. monocytogenes		< 2.5	> 0.92			500		Smelt and Hellemons(1998)
Listeria innocua	Eggs	3	0.768			450	T = 20 °C	Ponce and others(1998)
L. monocytogenes	Ground Pork	1.89 - 4.17	0.552 - 1.219			414	T = 25 °C	Murano and others (1999)
L. monocytogenes	Ground Pork	0.37 - 0.63	3.656 - 6.224			414	T = 50 °C	Murano and others (1999)
L. monocytogenes ScottA		4	0.576			345	T = 50 °C	Kalchayanand et al. (1998)
P.fluorescens	Milk	1	2.303			300	T = 50 °C	Gervilla et al. (1997b)
P. fluorescens	Milk	1	2.303			400	T = 10 °C	Gervilla et al.(1997b)
P. fluorescens		0.6	3.838			345	T = 50 °C	Kalchayanand et al.(1998)
Spores								
Clostridium sporogenes		12	0.192			680		Crawford and others (1996)
C. sporogenes		16.772	0.138			600	T = 90 °C	Rovere et al.(1996b)
C. sporogenes		6.756	0.341	725 (90 °C)		700	T = 93 °C	

C. sporogenes		5.306	0.434			800	T = 93 °C	
C. sporogenes		3.502	0.658			600	T = 100 °C	Rovere et al. (1996b)
C. sporogenes		3.186	0.723	962 (100 °C)		700	T = 100 °C	
C. sporogenes		2.857	0.806			800	T = 98 °C	
C. sporogenes		1.282	1.796			600	T = 108 °C	Rovere et al. (1996b)
C. sporogenes		0.901	2.556	752 (108 °C)		700	T = 108 °C	
C. sporogenes		0.695	3.314			800	T = 108 °C	
C. botulinum-TypeE Alaska	Buffer	8.77	0.263			758	T = 35 °C	Reddy and others (1999)
C. botulinum TypeE Alaska	Buffer	2.64	0.872			827	T = 35 °C	
C. botulinum TypeE Beluga	Crab Meat	3.38	0.681			758	T = 35 °C	Reddy and others (1999)
C. botulinum TypeE Beluga	Crab Meat	1.64	1.404			827	T = 35 °C	
C. botulinum-TypeE Alaska	Crab Meat	2	1.152			758	T = 35 °C	Reddy and others (1999)
C. botulinum TypeE Alaska	Crab Meat	1.76	1.309			827	T = 35 °C	
C. botulinum TypeA 62-A	Buffer	13.21	0.174			414	T = 75 °C	
C. botulinum TypeA 62-A	Buffer	12.6	0.183			551	T = 75 °C	
C. botulinum TypeA 62-A	Buffer	10.59	0.218	1524	4.39×10^{-6}	689	T = 75 °C	Reddy and others (1999)
C. botulinum TypeA 62-A	Buffer	9.19	0.251			758	T = 75 °C	
C. botulinum TypeA 62-A	Buffer	6.7	0.344			827	T = 75 °C	
Yeast								
Saccharomyces cerevisiae	Orange Juice	10.81	0.21			300	T = 34 °C	Zook and others (1999)
		2.8	0.82			350	T = 36.8 °C	
		0.97	2.37	117	1.24×10^{-4}	400		
		0.5	4.61			450		
		0.18	12.79			500		
S. cerevisiae	Apple Juice	9.97	0.231			300	T = 34 °C	Zook and others (1999)
		0.88	2.617	115	1.371×10^{-4}	400		

		0.28	4.798			450	T = 39.7 °C	
		0.15	15.35			500	T = 43.4C	
S. cerevisiae		1.27	1.813			350	pH = 3.7	Parsih et al. (1998)
S. cerevisiae		0.067	34.373			500	pH = 3.7	

Table 10-1c ***Kinetic Parameters for Inactivation of Microbial Population for Pulsed Electric Field Processes (PEF)***

Microorganism	Substrate	Time Parameter		PEF Coefficient		Field Strength	Other	Reference
		(D)	(k)	[Z(E)]	(K)			
		(microsec)	($x 10^{-2}$/ microsec)	(kV/ cm)	(kV/ cm)	(kV/cm)		
Vegetative Cells								
Escherichia coli	Skim Milk	38.4 - 44.8	5.14 - 6.0			20 -45	T = 15 °C	Martin Belloso et al. (1997b)
E. coli	SMUF	17.8	12.94			70	T = 20 °C	Zhang et al. (1995a)
E. coli	Milk	333	0.692			22		Grahl and others (1992)
E. coli	SMUF	4000 - 4500	0.051 - 0.058			16	T = 37 °C	Pothakmury et al. (1995)
E. coli	Buffer	75 - 100	2.3 - 3.07			20	T < 30 °C	Hulshergerand Nieman (1980)
E. coli	Phosphate Buffer	270	0.853	~41	6.3 - 8.1	20	T = 20 °C	Hulsheger et al. (1983)
E. coli	0.1% NaCl	100	2.3			19.5	T = 20 °C	Sale and Hamilton (1967)
E. coli	Phosphate Buffer	2	115.15			40		Matsumoto et al. (1991)
E. coli	Potato Dextrose	16 - 32	14.39			40	T = 15 °C	Zhang et al. (1994b)
E. coli	Skim Milk	64 - 96	2.4 - 3.6			40	T = 15 °C	Zhang et al. (1994b)
E. coli	Skim Milk	27.4 - 49.6	4.64 - 8.41			50	T < 30 °C	Qin et al. (1995c)
E. coli	SMUF	26.7	8.63			50	T < 30 °C	Qin et al. (1995c)
Salmonella Dublin	Skim Milk	4 - 42.4	0.054 - 0.52			15 - 40	T=15-40°C	Sensoy and others (1997)

S. Dublin	Milk	360	0.64			36.7	T = 63 °C	Dunn and Pearlman (1987)
Salmonella Typhimurium	NaCl	4	57.58			83		Gupta and Murray (1989)
L. monocytogenes (Scott A)	Milk	150-200	0.012-0.015			30	T = 10-50 °C	Reina and others (1998)
L. monocytogenes	Phosphate Buffer	540	0.426		6.4-6.5 (2- 2.8)	20		Hulsheger and others (1983)
Listeria innocua	Skim Milk	76.9	2.995			50	T = 15-28 °C	Fernandez and others (1999)
L. innocua	Skim Milk	26.7	8.625			50	T = 22-34 °C	Calderon-Miranda (1998)
L. innocua	Liquid Whole Egg	18.8	12.25			50	T = 26-36 °C	Calderon-Miranda (1998)
Staphylococcus aureus	SMUF	4000 - 6000	0.038-0.058			16	T = 37 °C	Hulsherger and others (1983)
S. aureus	SMUF	4000-4500	0.052 - 0.058			16	T > 30 °C	Pothakamury et al. (1995)
S. aureus	Phosphate Buffer	360	0.64		2.6 (2.0)	20		Hulsherger and others (1983)
Lactobacillus delbrueckii	SMUF	2000-2400	0.096 - 0.115			16	T < 30 °C	Polhakamury (1995)
Lactobacillus delbrueckii	Buffer	1022	0.225		(1.6)	25	T = 60 °C	Jayaram and others (1992)
Pseudomonas fluorescens	Skim Milk	22.2	10.374			50	T = 15-28 °C	Fernandez and others (1999)
Pseudomonas auriginosa	Phosphate Buffer	308.6	0.746		6.3 (1.8 - 2.6)	20		Hulsheger and others (1983)
P. fluorescens	Water	3.3	69.79			10	T = 20 °C	Ho and others (1995)
Klebsiella pneumoniae	Phosphate Buffer	360	0.64		6.6	20		Hulsheger and others (1983)
Spores								
Bacillus cereus	0.15% NaCl	50- 60	3.84-4.61			50	T = 25 °C	Marquez and others (1997)
Bacillus subtilis	0.15% NaCl	17.5-26.3	8.76-13.16			50	T = 25 °C	Marquez and others (1997)
B. subtilis	SMUF	2500-3000	0.077 - 0.092			16	T < 30 °C	Pothakamury (1995)
B. subtilis	Pea Soup	11.3	20.38	~15.5		33	T < 55.5 °C	Vega-Mercado et al. (1996a)
B. subtilis	SMUF	425-520	0.44-0.54			16		Qin and others (1994)

Yeast								
Saccharomyces cerevisiae	NaCl	61.5	3.745			35		Jacob et al. (1981)
S. cerevisiae	Phosphate Buffer	270	0.853		(2.3)	20		Hulsheger et al. (1983)
S. cerevisiae	Water	4666	0.049			20		Mizuno and Hori (1988)
S. cerevisiae	Potato Dextrose	8.7	26.47			40	T = 15 °C	Zhang et al. (1994b)
S. cerevisiae	Apple Juice	102.9-135	1.706 - 2.238	~17		12	T=4-10 °C	Zhang et al. (1994a)
S. cerevisiae	Apple Juice	42.9 - 428.6	0.537 - 5.368			25	T < 30 °C	Qin et al. (1995a)
S. cerevisiae	Apple Juice	0.83	277.47			50	T=22-29 °C	Qin and others (1995a)
Candida albicans	Phosphate Buffer	240	0.96		2.2 (1.2-3.1)	20		Hulsheger et al. (1983)
Zygosaccharomyces bailli	Juices	0.4-0.7	3.29-5.76			32 - 36.5	T = 20 °C	Raso et al. (1998)

Microbiological Critical Factors

The efficacy of a preservation technology is influenced by a number of microorganism-related factors that are generally independent of the technology itself. These include the type and form of target microorganism; the genus, species and strain of microorganism; growth stage; environmental stress selection mechanisms; and sublethal injury. Each of these factors influences the resistance of a microorganism to a preservation process, independently of the apparent inactivation capacity of that particular process.

Among the food microbial hazards, bacteria are generally the most resistant microorganisms of concern and therefore should be the primary targets in most preservation processes. In most cases, microorganisms other than bacteria will be destroyed before or concurrently with pathogenic and spoilage bacteria; however, in designing processes to inactivate all pathogens, it is also advisable to consider the resistance properties in foods of other microorganisms such as yeasts, molds, parasites and protozoa, that may persist in or grow in foods.

A few genera of foodborne bacteria (for example, *Clostridium spp.* and *Bacillus spp.*) are capable of existing in 2 forms: active vegetative cells and dormant spores. These 2 forms often differ in their resistance properties to heat, chemicals, irradiation and other environmental stresses. In that same manner, studies have shown that spores are typically more resistant than vegetative cells to the alternative processing technologies evaluated in this report. For pasteurization purposes, one is mostly concerned with the inactivation of vegetative cells of disease-producing microorganisms. In order to have a commercially sterile product, however, one must devise a process that inactivates all microbial spores (usually targeting spores of *C. botulinum*) capable of germinating and growing in the food under normal storage conditions.

Differences in resistance of microorganisms may be found not only between genera and species but also between strains of the same species. For instance, some bacterial strains with unique resistance to thermal inactivation, irradiation and high pressure processing have been identified. It is possible that, in the future, a pathogenic "super bug" would emerge. If this occurs, this pathogen should be considered a possible food safety hazard and the process would have to be redesigned to specifically inactivate it. Alternatively, if the "super bug" is not a pathogen or spoilage microorganism, it may be very useful as a possible surrogate during development and validation of a process. Another factor that can affect resistance of bacteria to preservation processes is stage of growth. It appears that cells in exponential or log phase of growth are generally less resistant than cells in stationary phase. The development of stress resistance proteins in stationary phase may be the cause of this phenomenon.

One of the basic principles of microbial genetics is that extreme environments that would kill most bacterial cells result in the selection of mutants resistant to severe conditions. These environmental conditions encountered by a population of cells may induce a stress "defense mechanism" in some of them. This selection process has been scientifically supported by studies suggesting that bacterial stress may induce hypermutability. Hypermutability would in turn lead to a microbial population of greater resistance. Therefore, the exposure of cells to some form of stress may induce and allow the survival of microorganisms with unusually higher durability to a given inactivation process.

Mazotta found that the heat resistance of acid- or salt-adapted, heat-shocked, or starved *E. coli* O157:H7 cells was higher than that of cells grown to exponential or stationary phase under optimum conditions. He suggested that it would be appropriate to use stress-inducing culture conditions when studying the thermal resistance of vegetative pathogens in specific products in order to add an extra factor of safety to the process.

Lou and Yousef determined that sublethal stresses to ethanol, acid, hydrogen peroxide, heat, or salt had variable effects on subsequent exposure of *L. monocytogenes* to lethal levels of the same stressors. For example, heat shocking increased the resistance of the microorganism to ethanol, hydrogen peroxide and salt, but not to acid. Davidson reviewed the impact of stress induction on resistance to food antimicrobials. He stated that resistance could be acquired through previous exposure or adaptation due to cross-protection from environmental or processing factors including stresses such as heat or acid.

A number of studies like the ones described have demonstrated the occurrence of stress-induced enhanced resistance to inactivation. The questions relative to process design and verification are:

(1) Are the microorganisms and food environments likely to be of the type involved in stress induction?

(2) Would stress induced resistance possibly occur at any point in the food processing operation?

(3) If it did, would it significantly impact the inactivation process leading to possible underprocessing?

Considering that in most food processing systems the design is to have microorganisms exposed only once to a stress-inducing factor (for example, heat, acid, antimicrobials and so on), the development of a resistant population is not likely to occur. One possible exception might be the case of previously processed material that is reprocessed into the streamline. In those cases, in-depth studies of the impact of stress induction during the processing are needed.

Another microbial-related factor that influences the effectiveness of a process is the susceptibility of the microorganism to cellular injury. The effectiveness of an inactivation process is often measured by enumerating any surviving organisms (using biological indicators or surrogates) in a selective medium. Under the best circumstances, a processed microorganism would be either viable or dead; however, inactivation often results in a continuum of effects with some degree of injury. Injured cells can be easily underestimated, resulting in misleading conclusions about the efficiency of the method.

The detection and enumeration of injured microorganisms require special procedures. Often, injury is identified when surrogate organisms are enumerated using a selective culturing medium (generally a medium containing a chemical inhibitor that allows growth only of the particular microorganism being enumerated) in contast to a nonselective medium. It is often desirable to use selective media in the field to ensure growth of only the surrogate microorganism, and not of background microflora.

The choice of "best" method to enumerate the test organism will largely depend on the experimental variables and the researcher's experience with field studies.

Pathogens of Public Health Concern

In the United States, foodborne diseases caused by microorganisms can be attributed primarily to pathogenic bacteria, enteric viruses and protozoa. The following bacteria are known to be responsible for causing foodborne disease: *Aeromonas hydrophila, Bacillus cereus, Campylobacter jejuni, Clostridium botulinum, Clostridium perfringens, pathogenic Escherichia coli, Listeria monocytogenes, Salmonella, Shigella, Staphylococcus aureus,* pathogenic *Vibrio spp.* and *Yersinia enterocolitica*. The viruses of concern in foods are Hepatitis A, Norwalk, Norwalk-like and Rotavirus (CDC, 2000). *Cryptosporidium parvum, Cyclospora cayetanensis, Giardia lamblia* and *Toxoplasma gondii* are all parasites of concern, in part because they produce resistant cysts. When exploring the new preservation technologies, their preservation level should be compared to that of classical pasteurization or commercial sterilization technologies. Therefore, in an attempt to determine the pathogens of greatest public health concern for new technologies, the resistance of pathogens to heat will be examined.

Vegetative Bacteria Inactivated by Cooking and Pasteurization

Salmonella Bacteria of the genus *Salmonella* is one of the most well-known and frequently encountered pathogens in foods. Approximately 2,200 serovars of *Salmonella enterica subsp. enterica* exist and can be isolated from meats, poultry, eggs, raw milk, water, fish, shellfish, feeds, fruits and vegetables. Because *Salmonella* serovars are natural contaminants of intestinal tracts of animals, birds and reptiles, they may contaminate food and equipment through secondary contamination. Therefore, inactivation of this pathogen through processing and avoidance of post-processing contamination is very important. The infectious dose of this microorganism can be very low: It has been demonstrated that it may take only one cell to cause a person to become ill. This makes effective pasteurization critical to produce safe food.

Inadequate heating of foods of animal origin or cross contamination are the primary vehicles for salmonellosis outbreaks. Meat and poultry (that is, beef, turkey, chicken and pork) and homemade ice cream (generally due to the use of raw eggs), fruits and vegetables and salads have been the most frequently reported items (CDC 2000). Raw or improperly pasteurized milk and eggs, as well as other foods have also been associated with salmonellosis. Inadequate cooking or processing, improper cooling, ingestion of raw products and cross contamination of foods after cooking seem to be the major sources of *Salmonella* serovars.

The maximum growth temperature for *Salmonella* serovars is 49.5 °C. The microorganism is considered to be sensitive to heat and is killed easily by pasteurization of milk equivalent to 71.7 °C for 15 s. The heat resistance of *Salmonella* serovars depends on factors such as serovar type, water activity, pH and heating medium. The heat resistance of serovars in various foods is shown in Table 10-2. *Salmonella* serovar Senftenberg is generally considered to be the most resistant strain (Tables 10-2 and 10-3). Therefore, any heat-resistant studies using *Salmonella* serovars should include serovar Senftenberg strain 775W, unless it is not relevant for the application. Then it would serve only as a point of reference for heat resistance.

Shigella Bacteria of the genus *Shigella*, the causative microorganism for shigellosis, is a member of the *Enterobacteriaceae*. It is a gram-negative, non-sporeforming non-motile rod. The organism has a growth range of 10 to 47.2 °C with an optimum of 37 °C. The 2 primary foodborne pathogens are *S. flexneri* and *S. sonnei*. The microorganism is carried by humans and primates and is spread to food by carriers and contaminated water. *Shigella* strains are not particularly heat resistant. Approximately 5 min at 63 °C inactivates most strains of *S. flexneri* and *S. sonnei*. The main foods implicated in outbreaks of *Shigella spp.* are salads and seafoods that become contaminated during handling by infected workers or by unclean and unsanitized food contact surfaces. Control of *Shigella* is best accomplished by hygiene, health education, water disinfection and sanitation along with mild heat treatment where necessary.

Pathogenic Escherichia coli *Escherichia coli* is a gram-negative, motile, facultative anaerobe non-sporeforming rod. The source of the microorganism is generally the gastrointestinal tract of warm-blooded animals but it can also be found in water. Five to 6 types of diarrheagenic *E. coli* are known today, including enteropathogenic, enterotoxigenic, enteroinvasive, enterohemorrhagic, enteroadherent and, enteroaggregative. These strains may cause neonatal, infantile, traveler's, or bloody diarrhea. Some produce toxins while others are invasive.

Enterohemorrhagic *E. coli* (EHEC) causes a sequela called hemolytic uremic syndrome. Foods implicated in outbreaks of EHEC include ground beef, roast beef, alfalfa sprouts, raw milk, apple cider, meat sandwiches, mayonnaise, lettuce and dry salami. Food process inactivation of this bacterium is best accomplished by:

1) Adequate cooking of all meat products to a center point temperature of 165-180 °F.

2) A professional sanitation program in place to inactivate the bacteria and prevent post-processing contamination or cross-contamination of processed food with raw product.

The heat resistance of *E. coli* is equivalent to or slightly lower than *Salmonella* serovars (Table 4). Therefore, a heat process sufficient to inactivate *Salmonella* serovars will also likely inactivate *E. coli.*

Yersinia enterocolitica *Yersinia enterocolitica* is a pathogen that causes a foodborne infection with an onset time of 3-7 d. The symptoms of yersiniosis include severe abdominal cramps which mimic appendicitis, watery diarrhea, vomiting and fever. Pork and pork products, milk and foods washed in contaminated water (for example, tofu) have all been implicated in outbreaks. Cross contamination can also cause a problem in ready-to-eat foods. *Yersinia enterocolitica,* as a psychrotroph can grow at refrigeration temperatures. In fact, cold storage can be selective for the microorganism. *Yersinia enterocolitica* has very low heat resistance in milk (See Table 10-5).

Vibrio Three species of *Vibrio* are potential pathogens in food: *V. cholerae, V. parahaemolyticus and V. vulnificus.* All are found in the marine environments and contaminate foods via contaminated water. Foods associated with *Vibrio spp.* foodborne infections include seafoods, raw vegetables, milk and inadequately sanitized water. Keeping raw product at low temperatures prior to processing helps to slow growth, and heating above 60 °C should easily inactivate this heat-sensitive organism (See Table 10-6).

Aeromonas hydrophila *Aeromonas hydrophila* has many pathogenic properties resembling V. vulnificus, that is, gastroenteritis in healthy individuals or septicemia in individuals with underlying chronic disease (for example, leukemia, carcinoma and cirrhosis) and those treated with immunosuppressive drugs or who are undergoing cancer chemotherapy or with impaired immune systems. Infections among healthy people are generally self-limiting whereas children are at the greatest risk.

Species of *A. hydrophila* are ubiquitous in freshwater environments. Although *A. hydrophila* can be isolated from a wide range of foods at the retail level, outbreaks are generally small. The organism is a well-established component of raw meat spoilage and is found on beef, pork, lamb, poultry, fish and shellfish. It is also a common component of raw milk and raw vegetables. *Aeromonas hydrophila* is eliminated by mild heat treatments (See Table 10-7).

Campylobacter jejuni *Campylobacter jejuni* is the leading cause of foodborne illness in humans in the United States (CDC 1999). The organism causes a diarrheal infection but can also have a more severe sequela known as Guillain-Barre Syndrome (0.2–2 cases/1000 cases of *C. jejuni*; paralysis, demyelination of nerves). The microorganism is generally unable to grow in foods; however, it often finds entry to food via human carriers or contamination. In heat-treated or dehydrated foods, contamination may not be a problem, but raw refrigerated foods of animal origin can be a source. The organism can be isolated from all common food animals and birds. Foods commonly associated with infection include raw milk, poultry, red meat and contaminated water sources. The microorganism is extremely sensitive to heat and would be inactivated at temperatures as low as 55 °C (See Table 10-8).

Listeria monocytogenes *Listeria monocytogenes* is the cause of a foodborne illness known as listeriosis. There are 13 serovars of pathogenic *L. monocytogenes*, but according to Raccourt and Cossart (1997), 95% of human isolates are serotypes 1/2a, 1/2b and 4b. The microorganism often attacks persons with suppressed immune systems, including pregnant women, neonates, the elderly and persons immunosuppressed by medications. Manifestations of listeriosis include abortion, perinatal septicemia and meningitis. The mortality rate associated with listeriosis is ca. 30% (Rocourt and Cossart 1997; V. J. Scott, personal communication).

Although rarely described, *L. monocytogenes* can grow to high populations in temperature-abused food resulting in severe gastroenteritis after consumption. *Listeria monocytogenes* is truly ubiquitous and can survive for long periods of time under extreme and adverse conditions. *Listeria monocytogenes* can multiply in foods stored at refrigeration temperatures so risk may increase during storage. It has been found in raw milk, raw milk cheese, soft-ripened cheeses, raw meats and seafood. There have been cases of illness from coleslaw and other raw vegetables that have been fertilized with animal manure or wastewater, and then not rinsed and cleaned prior to preparing and eating.

Recently, an outbreak of listeriosis caused by *L. monocytogenes* serotype 4b resulted in 50 cases and 8 deaths linked to consumption of hot dogs and/or deli meats (CDC 1998). *Listeria monocytogenes* may be controlled in food processing plants with a good sanitation program and prevention of cross-contamination between raw and finished product. In addition, heating equivalent to milk pasteurization (71.7 °C for 15 s) or heating to 62.8 °C for 30 min or above should inactivate the microorganism in milk or other foods (SeeTables 10-9 and 11-10).

Vegetative Bacteria Inactivated by Pasteurization But Able to Produce a Heat-stable Toxin

Staphylococcus aureus *Staphylococcus aureus* can produce a toxin in improperly stored food that, if ingested, will produce mild to severe symptoms of nausea, cramps, vomiting, diarrhea and prostration in 2-7 h, lasting 1 to 2 d. The enterotoxins produced by *S. aureus* are resistant to heating (up to a D110C = 10 min as measured by bioassays), including low-acid canned food processing. Many healthy people harbor *S. aureus*. It can be found in the nose, throat, hands, fingertips, hair and skin. Any food that is contaminated with the organism and supports growth can potentially develop this bacterial toxin.

Proteinaceous foods (for example, chicken, turkey, meat, fish, dairy products, salad vegetables, ham), potatoes, cream-filled products and cream sauces are commonly involved in outbreaks. As the microorganism does not compete well in mixed populations, it is generally not a problem in unheated foods; however, when other naturally occurring bacteria are killed by cooking or inhibited from growth and *S. aureus* is later introduced by humans, it survives and grows. Consequently, it may be found in prepared foods such as salads, custards and cream-filled products. *Staphylococcus aureus* is also resistant to low water activities and survives curing solutions that contain salt or sugar. The cells of the microorganism are not heat resistant (Table 11) and should easily be killed by any mild heat treatment. *Staphylococcus aureus* is best controlled by preventing contamination.

Sporeforming Bacterial Organisms That Survive Cooking and Pasteurization and Produce Toxin

Bacillus cereus *Bacillus cereus* is a sporeforming organism that produces 2 types of illness: the diarrheal syndrome, which develops within 20 h following ingestion, or the emetic (vomiting) response, which occurs 1 to 5 h after ingestion. The illnesses are the result of toxins associated with growth of the microorganism in foods (emetic) or the gastrointestinal tract (diarrheal). The diarrheal (enterotoxin) is produced during exponential growth in the gastrointestinal tract, while the emetic toxin is produced by cells growing in the food product (Granum, 1997).

Dairy products, cereals, meats and fried rice are commonly the foods involved in foodborne illness. Since the microorganism does produce spores, it is considered resistant to at least pasteurization conditions. In fact, the spores of *B. cereus* have D95C values ranging from 1.5 to 36.2 min (Table 10-12). One thing that is apparent concerning the heat resistance of *B. cereus* spores is how variable their heat resistance is in the same food product. It is generally thought that *B. cereus* do not survive the low-acid canned food process. For instance, using the highest z(T) (10.1) and D-value (36.2) in Table 10-12, the D121C is 5.8 s, which confirms that this heat susceptibility is the case (ICMSF 1996).

Clostridium perfringens *Clostridium perfringens* food poisoning is the result of an enterotoxin produced in the gastrointestinal tract by *C. perfringens.* After ingestion, the microorganisms multiply and sporulate in the small intestine, releasing the enterotoxin and causing symptoms including abdominal cramps and diarrhea (McLane 1997). Meat and poultry are the most common foods associated with *C. perfringens* food poisoning. Foodborne illness caused by this microorganism is generally the result of poor refrigeration and inadequate reheating of cooked foods.

Spores are often found on raw meats and may survive cooking of beef or poultry. If foods are inadequately cooled, the spores may then germinate and outgrow. If the food is then inadequately reheated, cells may continue to grow and reach large numbers. Ingestion of large numbers of organisms is necessary both for the microorganism to survive passage through the stomach and to initiate growth and toxin production in the intestines. As can be seen in Table 10.13, the D-value of *C. perfringens* spores at 98.9 °C may be as high as 31.4 min in beef gravy. Therefore, a low number of spores could potentially survive cooking of a meat in sauce; however, the preferred method of control of this microorganism is not necessarily initial heating but rather adequate cooling and adequate reheating following cooling to inactivate any cells produced during cooling.

Clostridium botulinum *Clostridium botulinum* is a common soil bacterium that produces heat-resistant spores. This organism produces a potent neurotoxin that may be toxic to both humans and animals. The toxin is considered heat-labile and can be inactivated by heating to 80 °C for 10 min. When ingested, the toxin is absorbed and irreversibly binds to peripheral motor nerves causing paralysis and possible death without antitoxin treatment. Vegetables can carry heat-resistant Type A, B and F *Clostridium botulinum* spores that are a major concern in low-acid canned foods. Type E spores also can be found in fish and seafood products. The heat resistance as measured by D-values of *C. botulinum* Types A and proteolytic B spores generally ranges from 0.6 to 3 min at 110 °C (Tables 10.14 and 10.15). A comparison of Type A and B spores in similar products is shown in Table 16. *Clostridium botulinum* Type E spores are much less resistant than Type A or B and can be inactivated at or below 100 °C (See Table 10-17).

Cyst-producing Protozoa That Can Remain Infectious in Unpasteurized Foods

***Cryptosporidium* and *Cyclospora*.** The protozoa, *Cryptosporidium parvum* and *Cyclospora cayetanensis*, are not able to replicate in foods, but they do produce cysts that can remain infectious in foods for extended periods of time. Since these organisms appear to have a low infective dose, their presence can contribute to infection, causing diarrhea in the general population.

These protozoa may contaminate water systems since the cysts have a high tolerance for disinfectants, such as chlorine. Washing food with contaminated water can infect foods with cysts. Thermal resistance values for *Cryptosporidium oocysts* of D60 °C of 20 s in distilled water and D71.7 °C of <1 s have been reported in milk. According to Rose and Slifko (1999), the heat resistance of *Cyclospora* may be similar to that of Cryptosporidium (See Table 10-18).

Enteric Viruses That Can Cause Foodborne Infection from Unpasteurized Foods

Hepatitis A virus, Rotavirus and Norwalk virus Several outbreaks of foodborne illness have been attributed to the viral contamination of shellfish and of unprocessed fruits. Hepatitis A virus and other enteric viruses may be found in shellfish taken from waters polluted by sewage. Fruits grown in fields where human waste or sludge is used as fertilizer have the potential for contamination by enteric viruses. Foods most vulnerable to viral contamination would be those not receiving a heat pasteurization step.

Surrogate Organisms to Validate Processing Parameters

Thermal, Electrothermal and Non-Thermal Food Processing

The establishment of traditional thermal processes for foods has been based on 2 main factors:

1) *Knowledge of the thermal inactivation kinetics of the most heat-resistant pathogen of concern for each specific food product.*

2) *Determination of the nature of heat transfer properties of the food system, generally defined by a heat transfer rate.*

These 2 factors are used to calculate the scheduled process, thereby ensuring inactivation of pathogen(s) in that product. The validity of the established process is often confirmed using an inoculated test pack study.

An inoculated pack study would be tested under actual plant conditions (this includes processing and control equipment, product and packaging) to reproduce the process in every detail. Since it is unwise to introduce viable pathogens into the production area, surrogate organisms are often utilized in the inoculated pack study, and their inactivation is measured to validate the process.

Surrogates play an important role as biological indicators that can mimic the thermal inactivation properties of a pathogen and can help to detect peculiarities or deviations in the processing procedure. One of the challenges in using new processing technologies for food preservation and pathogen inactivation is to determine if traditional methodologies can be used to establish and validate the new process. For practical purposes, the mechanism of microbial inactivation under electrothermal processes is basically the same as under conventional thermal processes: that is, heat inactivation.

Thus, the 2 factors described above, which are well established for thermal processes (Anonymous, 1996), should be used as a basis for establishing and validating scheduled electrothermal processes. It is also appropriate to use surrogate organisms to assist in determining and validating the process effectiveness. Regarding other preservation processes not based on heat inactivation (that is, high pressure, pulsed electric field, pulsed light), nonpathogenic surrogates still need to be identified and their significance evaluated. To accomplish this, more research needs to be done in the area of inactivation kinetics of pathogens by new technologies as well as in the identification of non-pathogenic candidates useful as surrogate organisms.

Importance of Surrogates

Surrogate organisms are invaluable in confirming the efficacy of thermal-based processes. Their use, as opposed to using actual pathogens, derives from the need to prevent the introduction of harmful organisms into the production facility area. The consequences of mishandling a pathogen in the presence of workers, product and equipment (from safety to legal liability issues) could be devastating.

Therefore, the use of surrogates by processing companies is of great importance to ensure microbiological safety of the process. For instance, surrogates have been used for many years in the low-acid canning industry to establish and validate the destruction of *C. botulinum* spores. The use of nonpathogenic spores of the putrefactive anaerobe *C. sporogenes*, or spores of the flat-sour thermophilic organism *Bacillus stearothermophilus* as surrogates for *C. botulinum*, have helped the industry develop processes that ensure products are safe and commercially sterile.

Criteria for Surrogates

The ideal surrogate would be the pathogen (or target organism) itself that had been transformed into a nonpathogenic form using genetic engineering techniques. Such an approach to surrogate selection is generally not followed due to possible reversion to pathogenicity or possible detection of false positives during routine testing. Generally, surrogates are selected from the population of well-known organisms that have well-defined characteristics and a long history of being non-pathogenic.

In selecting surrogates, the following microbial characteristics are desirable:

- *Nonpathogenic.*
- *Thermal or other inactivation characteristics that can be used to predict those of the target organism.*
- *Durability to food and processing parameters similar to target organism (for example, pH stability, refrigeration stability, oxygen tolerance and so on.)*
- *Stable thermal and growth characteristics that are similar from batch to batch.*
- *Easily prepared into high-density population forms.*
- *Once prepared, population is constant until utilized.*
- *Easily enumerated using rapid and inexpensive detection systems.*
- *Easily differentiated from other natural flora.*
- *Follows inactivation kinetics in a manner similar to pathogens receiving the same inactivation treatment.*
- *Genetically stable so results can be reproduced independently of laboratory or time of experiment.*
- *Will not establish itself as a "spoilage" organism on equipment or in the production area.*
- *Not susceptible to injury or reversible inactivation properties.*

- *Ideal surrogates, with all of the features described above, are scarce. Generally, surrogates will have many of the criteria, as is the case with the traditional surrogates used in low-acid canned foods processing validation.*

Surrogates for Pasteurized Products

Rather than using biological indicators as a basis for process validation, pasteurization processes have traditionally been evaluated and monitored using enzymatic destruction. Milk pasteurization has relied on the inactivation of the naturally occurring phosphatase enzyme as confirmation that product has received the proper heat treatment. This approach is not quantitative and is specific to pasteurization of milk.

To obtain quantitative information to support the development and validation of thermal processes, the use of bacterial surrogates is preferred to the use of naturally occurring enzymes. Research, however, is progressing on the identification and use of proteins and enzymes with inactivation kinetics comparable to microorganisms of concern, and there is a great potential in using them as chemical indicators.

Selection of surrogate organisms for pasteurization of products is a relatively new task for food scientists. The literature basically lacks information on recommendations for useful surrogate organisms. Therefore, a good deal of research and development may be required for progress to be made. Efforts are needed to identify and establish surrogates that meet many of the criteria listed above and can be utilized in process development for pasteurized foods.

Pathogens of public health significance in foods are vegetative cells of both gram-positive and gram-negative organisms as well as protozoan cysts. These organisms are inactivated fairly rapidly at pasteurization temperatures and are not good candidates as surrogates. Thermoduric lactic acid bacteria are also present in many foods. These organisms may survive the process and should be evaluated for their possible value as surrogates. They include the lactic streptococci (*Streptococcus thermophilus*), the lactobacilli (*Lacto-*

bacillus delbrueckii spp., *Lactobacillus bulgaricus,* and *Lactobacillus lactis*) and *Pedioccocus spp.* Other related lactic acid bacteria to consider as possible surrogates would include strains of *Lactococcus* and *Leuconostoc.* In cases where *Listeria monocytogenes* is the pathogen of interest, strains of *Listeria innocua* have served as non-pathogenic surrogates. In addition, nonpathogenic strains of *E. coli* have served as surrogates for *E. coli* O157:H7. In cases such as these, where surrogates are utilized, it should be proven that the surrogates are suitable for use based upon the above criteria.

Surrogates for Low-Acid Canned Food Products

Methodologies for validating low-acid-canned food (LACF) processes have been in existence for quite some time. The regulatory agencies require that scheduled processes for LACFs be established by a person or organization having expert knowledge of thermal processing requirements for foods packaged in hermetically sealed containers.

The biological validation of an electrothermal process such as ohmic heating or inductive heating may be designed and performed following conventional heating biological validation procedures; however, the mode of heat generation, heat distribution and location of coldest point(s) need special considerations when validating an electrothermal process and cannot be extrapolated from conventional thermal processes, as will be described in the subsequent sections of this report. These issues need to be investigated by experts in electrothermal processes.

Where novel non-thermal processes are being investigated, the application of traditional and classical approaches, such as the use of the formula described by C.O. Ball in the 1920's, may not be applicable. The approaches to process development for novel processes may initially be mostly empirical and strongly supported by biological validation until the nature of the kinetics of inactivation is more fully understood and hence predictable.

It remains to be seen if the use of organisms like *C. sporogenes* or *B. stearothermophilus* will be acceptable surrogates for *C. botulinum* in many of these novel non-thermal processing applications. As will be described in the following sections of this report, future research needs to address the resistance of pathogens of concern and the identification of appropriate surrogates for the specific non-thermal processes.

Other Considerations

The use of surrogate organisms to determine and validate processes for electrothermal processes will be challenging, especially for pasteurization processes. The following are some further points to consider while undertaking this endeavor:

- *Keep the approach as easy, accurate and simple as possible.*

- *Design the process so that the surrogate exhibits a predictable time-temperature profile that correlates to that of the target pathogen.*

- *Be attentive to the introduction of system modifications or variables, which could alter the heat-transfer properties of the food leading to inaccurate results (for example, thermocouple probes changing heating rates, nutrients added to the product for surrogate growth altering viscosity and so on).*

- *Validate the susceptibility or tolerance of a surrogate, if it is known.*

- *When first developing a process, working with a mixture of potential surrogate strains may be useful in narrowing the parameters toward establishing an effective process.*

Thermal Resistance Tables

Table 10-2 ***Heat Resistance of Serovars of Salmonella in Various Foods***
(Source: ICMSF 1996)

Serovar	Food	Temp. (°C)	D-value (min)	Approx. z-value (°C)
Senftenberg	Beef Boullion	65.5	0.66	
	Pea Soup	65.5	1.11	5.6
	Skim Milk	65.5	1.11	
	Milk	65.6	0.56	4.4
	Milk Chocolate	70	480	18.9
	Milk Chocolate	71	276	
Typhimurium	Milk Chocolate	70	1050	17.7
	Milk Chocolate	71	396	
	TSB[1] + 10% MS[2]	55.2	4.7	4.5
	TSB + 30% MS	55	11	4.6
	TSB + 42% MS	55.1	18.3	
	Ground Beef	57	2.13-2.67	
Eastbourne	Milk Chocolate	71	270	
None Specified	Ground Beef	57.2	4.2	

[1] Trypticase Soy Broth [2] Milk Solids

Table 10-3 ***Heat Resistance as Measured by the D-value at 65.6 °C (150 °F) and Approximate Z-values of Salmonella Serovars in Milk***
(Source: ICMSF 1996)

Serovar	$D_{65.6\,oC}$ (s)	Approx. z-value (°C)
Anatum	1.4	5.0
Binza	1.5	5.3
Anatum	1.4	5.0
Binza	1.5	5.3

Newbrunswick	1.3	4.5
Senftenberg	34.0	4.4
Tennessee	1.4	4.9

Table 10-4 ***Effect of Heating on Pathogenic and Non-pathogenic Strains of Escherichia coli***
(Source: ICMSF 1996)

Product	Strain	Temp (°C)	D-value (min)
Raw Milk	ATCC 9637 (NP)[1]	57.2	1.3
Chocolate Milk	ATCC 9637 (NP)	57.2	2.6
40% Cream	ATCC 9637 (NP)	57.2	3.5
Ice Cream Mix	ATCC 9637 (NP)	57.2	5.1
Skim Milk	O111:B4	55	5.5
Whole Milk	O111:B4	55	6.6
Ground Beef[2]	O157:H7	57.2	4.5
Ground Beef[3]	O157:H7	57.2	4.1-6.4
Ground Beef[2]	O157:H7	62.8	0.4
Ground Beef[3]	O157:H7	62.8	0.26-0.47

[1] NP = Non-pathogenic [2] Doyle and others (1984)
[3] Line and others (1991)

Table 10-5 ***Effect of Heating on Yersinia enterocolitica in milk; Where Measured, Z-value Ranged From 4.0–5.78 °C***
(Source: ICMSF 1996)

Temperature (°C)	D-value
51.7	23.4-29.9
55	1.8-2.2
57.2	4.6-14.7
58	1.4-1.8
60	0.067-0.51
62	0.15-0.19

62.8	0.012-0.96
65	0.028
68.3	0.09

Table 10-6 ***Heat Resistance of Vibrio Species***
(Source: ICMSF 1996)

Species	Product	Temp (°C)	D-value (min)	z-value (°C)
Parahaemo-lyticus				5.6-12.4
	Fish Homogenate	48	10-16	
	Clam Homogenate	55	0.02-0.29	
	Crab Homogenate	55	2.5	
Cholerae				
	1% Peptone	60	0.63	17
	Crab Homogenate	60	2.65	21
	Oyster Homogenate	60	0.35	
Vulnificus	Similar tu *V. para-haemolyticus*			

Table 10-7 ***Aeromonas Hydrophila Heat Resistance***
(Source: ICMSF 1996)

Product	Temp (°C)	D-value (min)	z-value (0 °C)
Raw Milk	48	3.3-6.2	5.2-7.7
Saline	48	2.2-6.6	5.2-7.7

Table 10-8 ***Campylobacter Jejuni Heat Resistance***
(Source: ICMSF 1996)

Product	Temp (°C)	D-value (min)
Skim Milk	50	1.3-5.4
	55	0.74-1.0

Ground Beef	50	5.9-6.3
	56	0.62-0.96
Lamb	50	5.9-13.3
	55	0.96-1.26
Cooked Chicken	55	2.12-2.25

Table 10-9 ***Resistance of Listeria Monocytogenes to Heat in Milk Products***
(Source: ICMSF 1996)

Products	Temp (°C)	D-value (min)
Raw Milk, Raw Skim Milk, Raw Whole Milk, Cream	52.2	24.08-52.8
	57.8	3.97-8.17
	63.3	0.22-0.58
	66.1	0.10-0.29

Table 10-10 ***Heat Resistance of Listeria Monocytogenes in Various Products at 60 °C***
(Source: ICMSF 1996)

Product	D60 °C - value (min)
Ground Meat	3.12
Ground Meat, cured	16.7
Fermented Sausage	9.2-11
Roast Beef	3.5-4.5
Beef	3.8
Beef Homogenate	6.27-8.32
Naturally Contaminated Beef	1.6
Weiner Batter	2.3
Chicken Leg	5.6
Chicken Breast	8.7
Chicken Homogenate	5.02-5.29

Carrot Homogenate	5.02-7.76
RANGE	1.6-16.7

Table 10-11 ***Staphylococcus Aureus Vegetative Cell Heat Resistance***
(Source: ICMSF 1996)

V x	Temp (°C)	D-value (min)	Notes
Milk	50	10	z = 9.5 °C
	55	3	
	60	0.9	
	65	0.2	
	70	0.1	
	75	0.02	
Meat Macerate	60	6.0	+ 500 ppm Nitrite
Pasta, Semolina-egg Dough	60	3	a_w = 0. 92
		5	a_w = 0. 87
		8	a_w = 0. 85
		12	a_w = 0. 83
		>40	a_w = 0. 80
		32	a_w = 0. 76
		22	a_w = 0. 61
Phosphate Buffer	60	2.5	pH = 6.5

Table 10-12 ***Bacillus Cereus Spores Heat Resistance in Various Media at 95 °C***
(Source: ICMSF 1996)

Heating medium	D95 °C (min)	z-value (°C)
0.06 M Phosphate, pH 7.0	2.6-21.7	10.0-10.1

0.05 M Phosphate, pH 7.0[1]	2.1-3.4	
Phosphate, pH 7.0	13	
Distilled Water	1.5-36.2	
Water	36	6.7-8.3
Infant Formula, pH 6.3	2.7-15.3	8.1-8.7
Milk	1.8-19.1	9.4
Bread	2.9-36.2	
RANGE	1.5-36.2	6.7-10.1
[1]Contains Sorbitol, Glycerol or NaCl		

Table 10-13 ***Heat Resistance of Clostridium Perfringens Spores (Source: ICMSF 1996)***

Product/ Heating Medium	Temperature (°C)	D-value (min)
Phosphate Buffer, pH 7.0	90	0.015-8.7
Phosphate Buffer, pH 7.0	98.9	18.6
	104.4	3.15
	110	1.29
Beef Gravy, pH 7.0	98.9	31.4
	104.4	6.6
	110	0.5

Table 10-14 ***Heat Resistance of Clostridium Botulinum Strain 62A (Type A) Spores at 110 °C (Source: ICMSF 1996)***

Product	D-value (min)	z-value (°C)
Asparagus, canned, pH 5.04	1.22	8.8
Asparagus, canned, pH 5.42	0.61	7.9
Corn, canned	1.89	11.6
Macaroni creole, pH 7.0	2.48	8.8
Peas, puree	1.98	8.3

Peas canned, pH 5.24	0.61	7.6
Peas, canned, pH 6.0	1.22	7.5
Spanish Rice, pH 7.0	2.37	8.6
Spinach, canned, pH 5.37	0.61	8.4
Spinach, canned, pH 5.39	1.74	10.0
Squash	2.01	8.2
Tomato Juice, pH 4.2	1.50-1.59[1]	9.43
Tomato Juice, pH 4.2	0.92-0.98	-
Phosphate Buffer, M/15, pH 7.0	0.88	7.6
	1.74	10.0
	1.34	9.8
	1.6-1.9	8.1-9.2
	1.01	9.1
Distilled Water	1.79	8.5
[1]Strain A16037		

Table 10-15 ***Heat Resistance of Clostridium Botulinum Proteolytic Type B Spores at 110 °C***
(Source: ICMSF 1996)

Product	D-value (min)	z-value (°C)	Strain
Asparagus, canned, pH 5.04	1.09	9.7	213B
Asparagus, canned, pH 5.42	1.06	7.9	213B
Beans, snap	0.86	9.7	213B
Beets	1.17	10.8	213B
Carrots, fresh	0.94	9.4	213B
Corn	1.03	10.0	213B
Corn, puree	2.88	10.6	213B
Corn, canned	2.15	9.6	213B
Milk solids, whole, 20%, pH 6.34	0.93	7.9	213B
Mushrooms, puree	0.49-0.99	-	7 strains

Peas, puree	2.14-12.42	8.3	213B
Peas, canned, pH 5.6	3.07	10.1	213B
Peas, canned, pH 5.94	1.52	7.4	213B
Rock Lobster, liquor	2.97-3.33	10.6	A35
Spinach, fresh	1.75	10.3	213B
Spinach, canned, pH 5.39	1.54	8.6	213B
Spinach, canned, pH 5.37	1.19	-	213B
Phosphate Buffer, M/15, pH 7.0	1.85	7.7-11.3	213B
	1.4	8.5	213B
	1.6	8.3	213B
	1.19	9.1	213B
	2.0	9.1	213B
	1.5	10.1	Amanna
	2.0	9.0	169B

Table 10-16 ***Comparison of Heat Resistances as Measured by D-value in Minutes of Clostridium Botulinum Type A and B Spores in Similar Products at 110 °C***
(Source: ICMSF 1996)

Product	Type A	Type B (Proteolytic)
Asparagus, canned	0.61-1.22	1.06-1.09
Corn, canned	1.89	2.15
Peas, puree	1.98	2.14-12.42
Peas canned	0.61-1.22	1.52-3.07
Spinach, canned	0.61-1.74	1.19-1.54
Phosphate Buffer, M/15, pH 7.0	0.88-1.9	1.19-2.0

Table 10-17 ***Heat Resistance of Clostridium Botulinum Non-proteolytic Type E Spores in Seafood Products***
(Source: ICMSF 1996)

Product	Temp (°C)	D-value	z-value (°C)
Blue Crab Meat	74	6.8-13.0	-
Oyster Homogenate	70	72	7.5
Oyster Homogenate + 1% NaCl	70	100	6.8
Oyster Homogenate + 0.13% K sorbate	70	72	7.4
Oyster Homogenate + NaCl + K Sorbate	70	79	7.3
Menhaden Surimi	73.9	8.66	9.78

Table 10-18 ***Heating Resistance of Protozoa in Water Systems and in Foods***
(Sources: (1) ICMSF 1996); (2) Rose and Slifko, 1999)

Organism	Heating Med	Survival	Time/Temp.	Reference
Anisakis pseudo-terranova	Fish/ Medium	maximum survival time	1 min, 60 °C	1
Anisakis anisakis	Fish/ Medium	maximum survival time	10 s, 55 °C	1
Cryptosporidium oocysts	Distilled Water	3-log reduction	1 min, 60 °C	2
Cryptosporidium oocysts	Milk	6-log reduction	5 s, 71.7 °C	2
Taenia cysticerci		inactivated	60 °C	1
1 = ICMSF 1996 2 = Rose and Slifko 1999				

References

Alpas, H.; Kalchayandand, N.; Sikes, T.; Dunne, C. P.; and B. Ray. "Hydrostatic pressure among strains of foodborne pathogens." *Applied and Environmental Microbiology*. 65(9) (1994) : 4248-4251. Print.

Ananth, V.; Dickson, J. S.; Olson, D. G.; and E. A. Murano. "Shelf-life extension, safety and quality of fresh pork loin treated with high hydrostatic pressure." *J Food Protect*. 61(12) (1998) : 1649-1656. Print.

Anderson, W. A.; McClure, P. J.; Baird-Parker, A. C. and M. B. Cole. "The application of a log-logistic model to describe the thermal inactivation of clostridium botulinum at 213B at temperatures below 121-1 °C." *J Appl Bacteriol.* 80 (1996) : 283-290. Print.

Anonymous. *Canned Foods: principles of thermal process control, acidification and container closure evaluation.* Washington, DC. The Food Processors Institute. Fifth Edition. 1989. Print.

Anonymous. "Thermal processes for low-acid foods in metal containers." Washington, DC. *National Food Processors Association. Bulletin* 26-L (1996) : 72-85. Print.

Anonymous. US food & drug administration center for food safety and applied nutrition foodborne pathogenic microorganisms and natural toxins handbook (Bad Bug Book). 1999. http://vm.cfsan.fda.gov/list.html. Web. 2000.

Arabshahi, A.; and D. B. Lund. "Considerations in calculating kinetic parameters from experimental data." *J Food Process Eng.* 7 (1985) : 239-251. Print.

Buchanan, R. L. "Identifying and controlling emerging food-borne pathogens: research needs." *Emerging Infectious Diseases.* 3(4) (1997) : 517-521. Print.

Calderon-Miranda, M. L. *Inactivation of listeria inocua by pulsed electric fields and nisin.* Pullman, WA. Washington State University. 1998. Print.

Carsberg, H. "Sanitation: It's a small world." *Food Quality.* Jan/Feb (1999) : 61-62. Print.

CDC. "Update: Multistate outbreak of listeriosis-United States, 1998-1999." MMWR. 47 (1998) : 1117-1118.

CDC. "Incidence of foodborne illnesses: preliminary data from the foodborne diseases active surveillance network (Foodnet)-United States, 1998." MMWR. 48 (1999) : 189-194.

CDC. "Surveillance for foodborne disease outbreaks-United States, 1993-1997. Morbidity and Mortality Weekly Reports." 49 (.2000) : 1-51. Print.

Cho, H. Y.; Sastry, S. K. and A. E. Yousef. "Kinetics of inactivation of Bacillus subtilis spores by continuous or intermittent ohmic and conventional heating." *Biotechnol Bioeng.* 62(3) (1999) : 368-372. Print.

Crawford, Y. J., Murano, E. A., Olson, D. G. and K. Shenoy. "Use of high hydrostatic pressure and irradiation to eliminate Clostridium sporogenes in chicken breast." *J Food Protect.* 59 (1996) : 711-715. Print.

Datta, A. K. "Error estimates for approximate kinetic parameters used in food literature." *J Food Eng.* 18 (1992) : 181-199. Print.

Davidson, P. M. "Potential for emergence of resistance to antimicrobials used in the food industry." Proceeding Symposium on Food Microbiology. International Institute Life Science North America Tech Committe on Food Microbiology. p.19-21. Dearborn, MI.1999. Print.

Doyle, M. P.; and J. L. Schoeni. "Survival and growth characteristics of escherichia coli associated with hemorrhagic colitis." *Appl Env Microbiology*. 48 (1984) : 855-856. Print.

Dunn, J. E. and J. S. Pearlman. "Methods and apparatus for extending the shelf-life of fluid food products." Maxwell Laboratories, Inc. U. S. Patent 4,695,472. 1987. Print.

Fernandez-Molina, J. J.; Barkstrom, E.; Torstensson, P.; Barbosa-Canovas, G. V.; and B. G. Swanson. "Shelf-life extension of raw skim milk by combining heat and pulsed electric fields." Food Res. Int. 1999. Print.

Gervilla, R., Capellas, M., Ferragut, V. and B. Guamis. "Effect of high hydrostatic pressure on Listeria innocua 910 CECT inoculated into ewes' milk." *J Food Protect*. 60(1) (1997a) : 33-37. Print.

Grahl, T.; Sitzmann, W.; and H. Markl. "Killing of microorganisms in fluid media by high-voltage pulses." DECHEMA Biotechnology Conferences. 675-679. 1992. Print.

Granum, P. E. "Bacillus cereus." M. P. Doyle, L. R. Beauchat and T. J. 'Montville (eds.). *Food Microbiology: Fundamentals and Frontiers*. Washington, DC. American Society for Microbiology. 327-336. 1997. Print.

Gupta, R. P.; and W. Murray. "Pulsed high electric field sterilization." IEEE Pulsed Power Conf. Record. National Research Council. 58-64. 1988. Print.

Ho, S. Y.; Mittal, G.S.; Cross, J. D.; and M. W. Griffiths. "Inactivation of pseudomonas fluorescens by high voltage electric pulses." *J Food Sci*. 60(6) (1995) : 1337-1343. Print.

Hülsheger, H.; and E. G. Nieman. "Lethal effect of high-voltage pulses on e. coli K12." *Radiat Environ Biophys* 18(4) (1980) : 281-8. Print.

Hülsheger, H.; Pottel, J.; and E. G. Niemann. "Killing of bacteria with electric pulses of high field strength." *Radiat Environ Biophys.* 20 (1981) : 53-65. Print.

Hülsheger, H.; Pottel, J.; and E. G. Niemann. "Electric field effects on bacteria and yeast cells." **Radiat Environ Biophys**. 22 (1983) : 149-162. Print.

ICMSF. *Microorganisms in foods. Microbiological Specifications of Food Pathogens.* Book 5. Blackie Academic and Professional Publ. London. 1996. Print.

Jackson, G. J.; Leclerc, J. E.; Bier, J. W.; and; J. M. Madden. Cyclospora - still another new foodborne pathogen. *Food Technol.* 51(1) (1997) : 120. Print.

Jacob, H. E.; Forster, W.; and H. Berg. "Microbial implications of electric field effects. II. Inactivation of yeast cells and repair of their cell envelope." *Z Allg Microbial* 21(3) (1981) : 225-232. Print.

Jayaram, S.; Castle, G. S. P.; and A. Margaritis. "Kinetics of sterilization of Lactobacillus brevis cells by the application of high voltage pulses." *Biotechnol Bioeng.* 40(11) (1992) : 1412-1420. Print.

Jeyamkondan, S.; Jayas, D. S.; and R. A. Holley. "Pulsed electric field processing of foods: a review." *J Food Protect.* 62(9) (1999) : 1088-1096. Print.

Kalchayanand, N.; Sikes, A.; Dunne, C. P.; and B. Ray. "Interaction of hydrostatic pressure, time and temperature of pressurization and pediocin AcH on inactivation of foodborne bacteria." *Journal of Food Protection.* 61(4) (1998) : 425-431. Print.

Katsuyama, A. M. *The microbiology of sanitation. A. M. Katsuyama. Principles of Food Processing Sanitation.* Washington, DC. The Food Processors Institute. 72-84. 2nd Edition. 1993. Print.

Larkin, J. and N. R. Reddy. "Personal communication."1999. Print.

Line, J. E.; Fain, A. R.; Moran A. B.; Martin; L. M.; Lechowich, R. V.; Carosella, J. M.; and Brown, W. L. "Lethality of heat to escherichia coli O157:H7: D-value and z-value determinations in ground beef." *J Food Protect.* 54 (1991) : 762-766. Print.

Lou, Y.; and A. E. Yousef. "Adaptation to sublethal environmental stresses protects listeria monocytogenes against lethal preservation factors." *Appl Environ Microbiol.* 63(4) (1997) : 1252-1255. Print.

Marquez, V. O.; Mittal, G. S.; and M. W. Griffiths. "Destruction and inhibition of bacterial spores by high voltage pulsed electric field." *J Food Sci.* 62(2) (1997) : 399-401,409. Print.

Martin-Belloso, O.; Qin, B. L.; Chang, F. J.; Barbosa-Canovas, G. V.; and B. Swanson. "Inactivation of Escherichia coli in skim milk by high intensity pulsed electric fields." *J Food Process Eng.* 20 (1997b) : 317-336. Print.

Matsumoto, Y.; Satake, T.; Shioji, N.; and A. Sakuma, "Inactivation of microorganisms by pulsed high voltage applications." Conference Record of IEEE Industrial Applications Society Annual Meeting. 652-659. 1991. Print.

Mazotta, A. S. "Problems and solutions to development of pathogen resistance to thermal processing." ISSI Meeting. 1999. Print.

McLane, B. A. "Clostridium perfringens." M. P. Doyle, L. R. Beauchat and T. J. Montville (Eds.). *Food Microbiology: Fundamentals and Frontiers.* Washington, DC. American Society for Microbiology. 1997. Print.

Mead, P. S.; Slutsker, L.; Dietz, V.; McCaig, L. F.; Breese, J. S.; Shapiro, C.p Griffin, P. M.; and R. V Tauxe,. "Food-related illness and death in the United States." *Emerging Infectious Diseases.* 5 (1999) : 607-625. Print.

Metrick, C.; Hoover, D. G.; and D. F. Farkas. "Effects of high hydrostatic pressure on heat-resistant and heat-sensitive strains of Salmonella." *J Food Sci.* 54 (1989) : 1547-1564. Print.

Mizuno, A. and Hori, Y. "Destruction of living cells by pulsed high-voltage application." IEEE *Trans Ind Appl.* 24(3) (1998) : 387-394. Print.

Mussa, D. M.; Ramaswamy, H. S.; and J. P. Smith. "High-pressure destruction kinetics of listeria monocytogenes on pork." *J Food Protect.* 62(1) (1999) : 40-45. Print.

Patterson, M. F.; Quinn, M.; Simpson, R.; and A. Gilmour. "Sensitivity of vegetative pathogens to high hydrostatic pressure treatment in phosphate buffered saline and foods." *J Food Protect.* 58(5) (1995) : 524-529. Print.

Patterson, M. F. and D. J. Kilpatrick. "The combined effect of high hydrostatic pressure and mild heat on inactivation of pathogens in milk and poultry." *J Food Protect.* 61(4) (1998) : 432-436. Print.

Peleg, M. "A model of microbial survival after exposure to pulse electric fields." *Journal Sci Food Agric.* 67(1) (1995) : 93-99. Print.

Peleg, M.; and M. B. Cole. "Reinterpretation of Microbial Survival Curves." *Crit Rev in Food Science.* 38(5) (1998) : 353-380. Print.

Ponce, E.; Pla, R.; Mor-Mur, M.;Gervilla, R. and B. Guamis. "Inactivation of Listeria innocua inoculated in liquid whole egg by high hydrostatic pressure." *J Food Protect.* 61(1) (1998) : 119-122. Print.

Pothakamury, U. R.; Monsalve-Gonzalez, A.; Barbosa-Canovas, G. V.; and B. G. Swanson. "Inactivation of Escherichia coli and Staphylococcus aureus in model foods by pulsed electric field technology." *Food Res Int.* 28(2) (1995a) : 167-171. Print.

Qin, B.-L. "Nonthermal inactivation Saccharomyces cerevisiae in apple juice using pulsed electric fields." *Lebensm Wiss Technol.* 28(6) (1995) : 564-568. Print.

Qin, B.-L.; Chang, F.-J.; Barbosa-Canovas, G. V.; and B. G. Swanson. "Nonthermal inactivation of S. cerevisiae in apple juice using pulsed electric fields." *LebensmWiss Technol.* 28(6) (1995a) : 564-568. Print.

Raso, J., Pagan, R., Condon, S. and F. J. Sala. "Influence of temperature and pressure on the lethality of ultrasound." Appl Environ Microbiol. 64(2) (1998a) : 465-471. Print.

Reddy, N. R.; Solomon, H. M.; Fingerhut, G.; Balasubramaniam, V. M.; and E. J. Rhodehamel. "Inactivation of Clostridium botulinum types A and B spores by high-pressure processing." IFT Annual Meeting: Book of Abstracts. National Center for Food Safety and Technology, Illinois Inst of Technology. p. 33 Chicago, IL. 1999. Print.

The first edition of
Emerging Non-thermal Food Processing Technologies
By Dr. Perkins Muredzi
was produced by the publishing house
CBH Books
in Lawrence, Massachusetts
and printed in USA 2012.
For comments about this publication
or permissions requests, please write to:
CBH Books
A division of Cambridge BrickHouse, Inc.
60 Island Street
Lawrence, MA 01840 USA.

The publishing house,
CBH Books,
a division of Cambridge BrickHouse, Inc.
was created to encourage excellence in literature.
We publish works in all genres, in all languages,
in the U. S. and abroad.
For more information on how to publish with CBH Books,
please visit our website:

www.cbhbooks.com

Printed in Dunstable, United Kingdom

91336851R00156